Chapter 1: Introduction to AI-Driven Electrolysis

Overview of AI's Role in Modern Technology and Energy Solutions

Artificial Intelligence (AI) has emerged as one of the most transformative forces in the modern world. From revolutionizing industries like healthcare, finance, and entertainment to advancing fundamental scientific research, AI is continuously reshaping the way we live, work, and think. One of its most exciting applications is in the field of energy solutions, where AI is driving innovations that promise to change the way we generate, distribute, and consume power.

At its core, AI is about creating machines that can learn from data and make decisions with minimal human intervention. By analyzing vast amounts of information, AI can optimize systems, predict future trends, and enhance decision-making in ways that were previously unimaginable. In energy systems, AI can monitor energy consumption patterns, improve energy efficiency, and even predict equipment failures before they happen. More importantly, AI is making possible new forms of energy production that were once considered science fiction. Among these is AI-driven electrolysis, a technology that could not only revolutionize the way we produce hydrogen but also potentially unlock new sources of clean, sustainable energy for the world.

Introduction to Electrolysis and Its Significance in Hydrogen Energy

Electrolysis is a process that has been used for over a century to separate water into its constituent elements—hydrogen and oxygen—using electricity. When an electric current passes through water, it splits into hydrogen gas (H_2) and oxygen gas (O_2). This process is the basis for producing hydrogen, a fuel that can be used in various energy applications, including power generation, transportation, and industrial processes.

Hydrogen is considered one of the most promising clean energy carriers because it produces only water vapor when it is burned or used in fuel cells. Unlike fossil fuels, hydrogen is not associated with harmful emissions, making it an attractive alternative for reducing global carbon footprints and mitigating climate change. However, the widespread use of hydrogen has been limited by several challenges, including the high cost of production, storage, and distribution. Electrolysis has the potential to address these issues by providing a cleaner, more efficient way to generate hydrogen from renewable sources of energy.

For electrolysis to play a central role in the global energy transition, it must become more efficient and cost-effective. This is where AI comes in. By integrating AI with electrolysis systems, we can unlock new levels of efficiency, optimize the process, and reduce the costs associated with hydrogen production. AI can analyze real-time data from electrolysis operations, adjust variables like temperature and pressure, and predict the most efficient operating conditions. This dynamic optimization can make electrolysis a more viable and scalable technology for mass hydrogen production.

Why AI and Electrolysis Are Crucial in Shaping the Future of Human Energy Systems

The intersection of AI and electrolysis represents a paradigm shift in the way we think about energy. As the world grapples with the dual challenges of climate change and energy security, the need for innovative, sustainable energy solutions has never been more urgent. Hydrogen energy, when produced efficiently and sustainably, has the potential to serve as a cornerstone for a low-carbon future. However, to realize this potential, we must overcome significant hurdles in terms of cost, efficiency, and scalability.

AI offers a powerful tool for addressing these challenges. By leveraging machine learning algorithms, AI can optimize every aspect of the electrolysis process, from energy input to hydrogen output. This optimization can significantly reduce the energy required to produce hydrogen, making it more affordable and accessible. Additionally, AI can be used to integrate electrolysis with other renewable energy sources, such as solar and wind, to create a more resilient and sustainable energy infrastructure.

The ability to produce hydrogen on-demand, at scale, and with minimal environmental impact could transform industries across the globe. From transportation (where hydrogen fuel cells can power everything from cars to trucks to trains) to heavy industry (where hydrogen can be used as a clean feedstock for manufacturing processes), the applications for hydrogen are vast. With AI-driven electrolysis systems, we can make these applications more feasible and scalable, ultimately paving the way for a more sustainable energy future.

Moreover, AI's potential to optimize and control electrolysis systems could lead to entirely new possibilities for human energy use. Imagine a future where individuals, communities, and even entire cities can generate and store hydrogen energy using locally sourced water, all managed by AI systems that maximize efficiency and minimize waste. The integration of AI with electrolysis could create decentralized, self-sustaining energy networks that empower people to take control of their energy needs while contributing to the global effort to combat climate change.

In this book, we will explore how AI-driven electrolysis could unlock the potential of hydrogen as a clean energy source and how this technology could be integrated into human systems. From understanding the chemistry of electrolysis to examining the ethical considerations of using AI in human biology, we will delve into the science, possibilities, and challenges of this exciting frontier in energy innovation. The future of hydrogen energy lies in the hands of those who are pioneering the integration of AI with electrolysis, and this book will serve as a roadmap to understanding and realizing that future.

As we embark on this journey, it's important to recognize that while the road to AI-driven electrolysis and hydrogen energy may be complex, it is one filled with immense potential. By harnessing the power of AI, we can unlock new ways of producing, storing, and utilizing hydrogen energy, transforming the global energy landscape in the process.

Chapter 2: Understanding Electrolysis: The Science Behind the Process

Electrolysis, a fundamental chemical process, plays a critical role in the extraction of hydrogen from water—a key reaction in energy production. To truly grasp the potential of AI-driven electrolysis, it is essential to first understand the science that powers this process. This chapter will explore the basics of electrolysis, the history behind its discovery and development, and its significance in the context of hydrogen energy production.

What is Electrolysis?

At its core, electrolysis is a chemical reaction driven by an electrical current. The term "electrolysis" comes from the Greek words *electro-* (meaning "electricity") and *lysis* (meaning "separation"). The process involves the breakdown of compounds by an electric current. The most common application of electrolysis is the decomposition of water (H_2O) into hydrogen gas (H_2) and oxygen gas (O_2), which is vital for producing hydrogen energy.

The process of electrolysis requires two electrodes: an anode (positive electrode) and a cathode (negative electrode), immersed in an electrolyte, often water with a small amount of salt or acid to improve conductivity. When a voltage is applied across the electrodes, water molecules at the cathode are reduced (gaining electrons), producing hydrogen gas. Simultaneously, at the anode, water molecules are oxidized (losing electrons), releasing oxygen gas.

The overall reaction for the electrolysis of water is as follows:

$$2H_2O(l) \rightarrow 2H_2(g) + O_2(g)$$

This process, though seemingly simple, unlocks vast potential for energy production, especially in the context of renewable energy and hydrogen fuel.

The History of Electrolysis

The history of electrolysis dates back to the early 19th century. The process was first discovered in 1800 by English scientist **William Nicholson** and his colleague **Anthony Carlisle**, who used electrolysis to separate water into hydrogen and oxygen. Their work laid the foundation for what would eventually become the hydrogen economy.

Over the following decades, electrolysis was further developed and refined, gaining importance as a method for producing hydrogen in various industrial applications. The discovery of electrolysis also coincided with the rise of the industrial revolution and the development of technologies that could harness the power of electricity.

In the 20th century, electrolysis became a critical method for hydrogen production in industries such as fertilizers (for ammonia production) and petroleum refining. However, it wasn't until the global energy crisis of the 1970s and the growing demand for clean energy that electrolysis gained renewed interest, particularly in the context of producing hydrogen for use as a clean fuel.

The Role of Water in Generating Hydrogen

Water, H_2O, is a perfect starting material for the electrolysis process due to its molecular simplicity and abundance. The potential of water as a source of hydrogen has made it the cornerstone of the hydrogen economy. The electrolysis of water is seen as a way to unlock the full potential of hydrogen as an energy carrier, providing a clean and renewable source of energy.

Hydrogen, produced through water electrolysis, is considered one of the cleanest forms of energy. When used in fuel cells or combustion engines, hydrogen combines with oxygen to form water, releasing energy in the process without emitting any harmful pollutants or greenhouse gases. This makes it a perfect alternative to fossil fuels in the quest for sustainable energy solutions.

The role of water in hydrogen production also makes electrolysis a potentially renewable process. As long as there is access to water and electricity, hydrogen can be generated, making it a versatile and scalable solution. However, for electrolysis to be truly sustainable, the electricity used in the process must be derived from renewable sources, such as wind, solar, or hydroelectric power.

Electrolysis and its Conventional Uses in Energy Production

Electrolysis has been employed in various industrial applications for over a century. Traditionally, it has been used for the production of metals like aluminum and chlorine. In the context of hydrogen production, electrolysis has long been considered a clean, though energy-intensive, method for extracting hydrogen from water. The main traditional uses of electrolysis in energy production include:

1. **Hydrogen Production for Industrial Use**: Electrolysis is used in industries requiring large quantities of hydrogen, such as ammonia production for fertilizers and refining crude oil. While other methods like steam methane reforming (SMR) have been more widely used for hydrogen production, electrolysis offers a cleaner alternative when powered by renewable energy sources.

2. **Hydrogen for Fuel Cells**: The electrolysis process is key to the hydrogen fuel cell industry. Fuel cells convert hydrogen into electricity through an electrochemical process, with water as the only byproduct. As demand for clean energy solutions grows, hydrogen fuel cells have become increasingly important in transportation (hydrogen-powered cars), stationary power generation, and backup energy systems.

3. **Energy Storage**: Electrolysis also plays a role in the energy storage sector. When there is excess electricity available—such as during periods of high renewable energy production—water electrolysis can be used to produce hydrogen, which can then be stored and used later when energy demand exceeds supply. This is particularly important for renewable energy sources like solar and wind, which are intermittent by nature.

Why Electrolysis Matters for Hydrogen Energy

The electrolysis of water is a fundamental process that has enormous potential for the future of hydrogen energy. Hydrogen, as a clean and renewable fuel, could revolutionize how we power everything from cars to power plants. However, challenges exist in making electrolysis a viable large-scale solution for hydrogen production.

One of the primary challenges is the energy required to split water molecules. Electrolysis is an energy-intensive process, and the cost of producing hydrogen through electrolysis is currently higher than other methods, such as SMR. This makes it crucial to improve the efficiency of electrolysis systems and to ensure that the electricity used in the process comes from low-carbon sources.

As renewable energy technologies continue to advance, and as AI and automation optimize processes like electrolysis, the economics of hydrogen production will improve. In particular, AI could play a pivotal role in making electrolysis more efficient and cost-effective by managing the complex variables involved in the process, such as voltage, current, and pressure, which influence the efficiency of hydrogen production.

The Road to AI-Driven Electrolysis

AI has the potential to revolutionize the electrolysis process. Through machine learning, AI systems could optimize the operational parameters of electrolyzers, predict when maintenance is needed, and ensure that energy is used in the most efficient way possible. AI could also assist in improving the design of electrolyzers, finding new materials that improve efficiency, and even simulating electrolysis reactions to better understand how to achieve optimal results.

The combination of AI and electrolysis could make hydrogen production more affordable and scalable, accelerating the transition to a clean energy future. The application of AI could help solve many of the current limitations of electrolysis, from reducing energy consumption to increasing the efficiency of the system, potentially unlocking a new era of hydrogen energy.

Conclusion

In this chapter, we've explored the science of electrolysis, its historical development, and its current role in energy production. The process of electrolysis holds the key to unlocking hydrogen as a sustainable and clean energy source. However, the next frontier lies in enhancing the efficiency and scalability of this process through innovative technologies, particularly AI. As we move forward, understanding the fundamentals of electrolysis will be essential in shaping a future where AI and electrolysis play pivotal roles in human-centered energy systems. The following chapters will explore how these technologies can work together to create a new paradigm in energy production and human integration.

Chapter 3: Hydrogen Energy: A Path to Sustainable Power

In the search for sustainable energy solutions, hydrogen has emerged as a powerful candidate. This chapter explores the importance of hydrogen as a clean energy source, its applications across industries, and the challenges and opportunities associated with scaling hydrogen as a solution to the world's energy needs. By unlocking the potential of hydrogen, particularly through AI-driven electrolysis, we can address some of the most pressing energy challenges facing the globe.

The Importance of Hydrogen as a Clean Energy Source

Hydrogen is one of the most abundant elements in the universe and, when used as an energy carrier, it holds enormous potential for reducing our dependence on fossil fuels. Unlike traditional energy sources, hydrogen produces zero harmful emissions when it is used in fuel cells or combusted. The only byproduct of hydrogen fuel cells is water, making it an incredibly clean energy source. This contrasts sharply with the burning of fossil fuels, which releases carbon dioxide (CO_2) and other pollutants that contribute to climate change and air pollution.

As the world moves toward decarbonization, hydrogen offers a versatile solution. It can be used to power vehicles, industries, and homes, and can even serve as a storage medium for excess renewable energy. The versatility of hydrogen is what sets it apart from other clean energy sources, making it a key component of the energy transition.

Hydrogen also offers unique advantages over electricity in certain applications. For example, hydrogen has a much higher energy density than batteries, making it a more suitable energy carrier for long-duration storage and for use in heavy-duty transportation, such as trucks, buses, and even ships. The use of hydrogen in these applications could significantly reduce carbon emissions in sectors that are difficult to electrify.

Applications of Hydrogen Energy Across Industries

1. **Transportation** One of the most promising applications of hydrogen is in the transportation sector. Hydrogen fuel cells, which convert hydrogen into electricity to power electric motors, are already being used in vehicles like buses, trucks, trains, and even airplanes. Unlike battery electric vehicles (BEVs), which require long charging times and have limited range, hydrogen-powered vehicles can refuel quickly and travel longer distances.

   Hydrogen fuel cells are also seen as a viable option for heavy-duty and long-haul transportation, such as freight trucks and buses, which require large amounts of power and long operating hours. Major companies such as Toyota, Hyundai, and Nikola are already developing hydrogen-powered trucks, demonstrating the commercial viability of hydrogen in transportation.

2. **Industry and Manufacturing** Hydrogen is already used in several industrial processes, particularly in the production of ammonia for fertilizers, refining petroleum, and in the chemical industry for producing hydrogenated fats and oils. However, these processes are typically powered by hydrogen produced from fossil fuels, which release CO_2 into the atmosphere.

   By switching to hydrogen produced through electrolysis (using renewable energy), industries can significantly reduce their carbon footprint. This is particularly important for sectors like steel manufacturing, where hydrogen can be used as a cleaner alternative to coke (a carbon-rich material) in the production of steel.

3. **Energy Storage** One of the key challenges of renewable energy sources like solar and wind is their intermittency—energy is only generated when the sun shines or the wind blows. Hydrogen offers a solution to this problem through a process known as power-to-gas (P2G), where excess renewable electricity is used to produce hydrogen through electrolysis. This hydrogen can be stored and then used to generate electricity when demand is high or when renewable energy production is low.

   Hydrogen storage offers a unique advantage over batteries in terms of duration. While batteries are ideal for short-term storage (a few hours), hydrogen can be stored for much longer periods—months, or even years—making it a reliable medium for seasonal energy storage. This ability to store energy over long periods makes hydrogen a key enabler of a stable, renewable-powered energy grid.

4. **Heating and Electricity Generation** Hydrogen can also be used directly in heating systems or combined with natural gas in power plants. In the UK, for example, there are pilot projects underway to inject hydrogen into the existing natural gas infrastructure to reduce carbon emissions. Additionally, hydrogen can be used in gas turbines to generate electricity, either in dedicated hydrogen plants or in hybrid systems that use both hydrogen and natural gas.

While hydrogen-powered heating systems are still in the early stages of development, they offer a promising alternative to natural gas, particularly in areas where electrification is challenging due to the existing infrastructure.

Challenges in Scaling Hydrogen as an Energy Solution

While hydrogen offers many advantages, there are still several challenges that must be overcome to scale it as a mainstream energy solution:

1. **Cost of Production** The current cost of producing hydrogen through electrolysis is relatively high compared to other methods, such as steam methane reforming (SMR), which is the most common way of producing hydrogen today. However, SMR relies on natural gas and emits CO_2, while electrolysis powered by renewable energy is a much cleaner option.

   The key challenge is making green hydrogen (hydrogen produced through electrolysis using renewable energy) more cost-competitive. This requires reducing the cost of renewable electricity, improving the efficiency of electrolyzers, and scaling up production to drive down costs. As renewable energy prices continue to fall and new technologies emerge, the cost of green hydrogen is expected to become more competitive.

2. **Infrastructure Development** Another major barrier to hydrogen adoption is the lack of infrastructure for its production, storage, distribution, and use. Building a hydrogen economy requires substantial investment in new infrastructure, such as hydrogen refueling stations, pipelines, and storage facilities.

   In addition, the transportation and storage of hydrogen present unique challenges. Hydrogen is the smallest and lightest molecule, making it difficult to store and transport efficiently. Innovations in hydrogen storage technologies, such as advanced materials and liquefied hydrogen, are critical to making hydrogen a viable energy carrier on a large scale.

3. **Energy Density and Storage** While hydrogen has a high energy density by weight, it has a low energy density by volume, which makes it challenging to store in large quantities. Current hydrogen storage systems require high-pressure tanks or cryogenic temperatures to store hydrogen in a dense form. New storage methods, such as solid-state hydrogen storage or chemical hydrogen storage, are being researched to address this issue.

4. **Public Perception and Safety** Hydrogen is highly flammable, which can lead to safety concerns among the public. Although hydrogen has been used safely for many decades in industrial applications, ensuring the safe use of hydrogen in everyday applications requires careful planning, regulation, and public education. Safety standards for hydrogen infrastructure and fuel cells must be developed and adhered to, and risks associated with hydrogen leaks, combustion, and storage must be minimized.

Opportunities in Scaling Hydrogen

Despite these challenges, the potential for scaling hydrogen as a clean energy solution is enormous. Governments, industries, and research organizations worldwide are investing heavily in hydrogen research and infrastructure development. International initiatives, such as the European Union's Hydrogen Strategy and Japan's Hydrogen Roadmap, are accelerating the transition to a hydrogen economy.

Advances in AI, machine learning, and automation also offer exciting opportunities to optimize hydrogen production and storage. AI-driven electrolysis, for example, could enhance the efficiency of electrolyzers, reduce energy consumption, and improve the scalability of hydrogen production.

The growing focus on decarbonizing the energy sector, particularly in hard-to-abate sectors such as heavy industry, transportation, and aviation, presents a unique opportunity for hydrogen to play a central role in global energy systems. As the world moves towards a low-carbon future, hydrogen will be an essential component in achieving net-zero emissions.

Conclusion

Hydrogen energy holds great promise as a clean, sustainable, and versatile energy source. Its applications across industries such as transportation, energy storage, and manufacturing are vast, and its potential to decarbonize key sectors of the economy is unmatched. However, challenges remain in scaling hydrogen production, developing infrastructure, and improving cost efficiency.

AI-driven electrolysis offers a pathway to overcoming these challenges, providing a more efficient and cost-effective method of producing hydrogen. By combining advances in renewable energy, electrolysis, and AI, we can unlock the full potential of hydrogen as a clean energy solution, paving the way for a sustainable and hydrogen-powered future.

Chapter 4: The Rise of Artificial Intelligence in Energy Systems

As the world grapples with the challenge of transitioning from fossil fuels to renewable energy sources, artificial intelligence (AI) has emerged as a pivotal player in transforming energy systems. This chapter explores the increasing role of AI in energy production, how AI can optimize energy efficiency, and the potential for AI to revolutionize hydrogen energy production through electrolysis.

AI Applications in Various Energy Sectors

AI has already made significant strides in various energy sectors, revolutionizing the way we produce, store, and consume energy. Its applications span a wide range of fields, from electricity generation to energy storage and distribution, and even the integration of renewable energy sources into the grid.

1. **Smart Grids and Energy Distribution** AI's impact on energy distribution and management is perhaps most evident in the development of smart grids. A smart grid uses AI to optimize the generation, distribution, and consumption of electricity, ensuring that power is delivered efficiently and reliably. AI-enabled systems can predict electricity demand in real-time, balancing supply and demand while minimizing waste. This is particularly important for integrating renewable energy sources like wind and solar, which are intermittent and subject to fluctuation.

Smart grids powered by AI can also improve energy storage management, ensuring that excess energy generated during peak production times is stored efficiently for use during periods of low generation. This makes renewable energy more reliable and accessible, even when weather conditions are not favorable for energy production.

2. **Predictive Maintenance for Power Plants** In traditional energy systems, ensuring that power plants and energy infrastructure are running at peak efficiency is crucial for minimizing downtime and reducing operating costs. AI is being increasingly used in predictive maintenance, where machine learning algorithms analyze real-time data from power plants to predict when equipment will need maintenance or is at risk of failure. By identifying potential problems before they occur, AI allows operators to schedule repairs at the most convenient times, reducing the risk of unplanned shutdowns and extending the lifespan of equipment.

3. **Energy Demand Forecasting** Accurate forecasting of energy demand is essential for efficient energy production. AI-powered systems can predict energy usage patterns based on factors like weather, time of day, and historical usage data. This allows energy producers to adjust their generation and distribution strategies accordingly, ensuring that there is always enough energy available to meet demand without overproducing, which can waste resources and increase costs.

4. **Renewable Energy Integration** The integration of renewable energy into the grid is one of the most pressing challenges facing modern energy systems. Solar and wind power are variable by nature, making it difficult to predict their availability. AI can optimize renewable energy integration by forecasting the availability of solar and wind energy and adjusting grid operations to ensure that these sources are used as efficiently as possible. AI systems can also control the charging and discharging of energy storage systems, such as batteries, ensuring that excess renewable energy is stored and used when demand exceeds supply.

How AI Can Optimize Energy Production and Usage

AI offers unprecedented potential to optimize both energy production and consumption in ways that were previously impossible. By analyzing vast amounts of data, AI can provide insights that lead to significant efficiency gains in every stage of the energy production cycle.

1. **Optimizing Electrolysis for Hydrogen Production** In the context of hydrogen production, AI has the potential to revolutionize the electrolysis process. Electrolysis is an energy-intensive process, and its efficiency depends on numerous factors, including voltage, current, temperature, and pressure. AI can analyze these factors in real-time, adjusting the parameters to optimize energy consumption and maximize hydrogen output.

    Machine learning algorithms can also be trained to predict the most efficient operating conditions for electrolysis based on historical data and environmental factors. This level of optimization could significantly reduce the cost of producing green hydrogen, making it a more competitive alternative to hydrogen produced via traditional methods, such as steam methane reforming (SMR).

2. **Energy Efficiency in Buildings** AI-powered systems are increasingly being used in smart buildings to optimize energy usage. By collecting data from sensors embedded throughout the building, AI can adjust heating, cooling, lighting, and other systems to minimize energy waste while ensuring comfort for occupants. AI can also learn from historical data to predict energy usage patterns and optimize settings accordingly, reducing the building's overall energy consumption. Additionally, AI can help integrate renewable energy sources like solar panels into building energy systems, maximizing the use of on-site generation while reducing reliance on grid power. In this way, AI contributes to both energy efficiency and sustainability.

3. **Decentralized Energy Systems** As the energy landscape shifts towards more localized, decentralized systems, AI is playing a crucial role in enabling these transformations. Decentralized energy systems, such as microgrids and home energy management systems, use AI to optimize the generation, storage, and consumption of energy at the local level.

   These systems can integrate a variety of energy sources, including solar panels, wind turbines, and batteries, and use AI to determine the best way to balance energy production and consumption. For example, AI can decide when to store excess energy generated by solar panels for use during the night, or when to sell surplus energy back to the grid.

Potential for AI to Revolutionize Hydrogen Energy Production Through Electrolysis

The future of hydrogen as a clean energy source depends on the ability to produce it efficiently and cost-effectively. Hydrogen produced through electrolysis, using renewable electricity, is widely considered the "green" solution to hydrogen production. However, the cost of electrolysis remains relatively high due to the inefficiencies in current systems. AI offers a promising path toward reducing these inefficiencies and making hydrogen production more competitive.

1. **Real-time Monitoring and Control** AI can monitor the electrolysis process in real-time, adjusting parameters like temperature, voltage, and current to maximize efficiency. By doing so, AI can help minimize energy consumption and ensure that the system operates at optimal performance. AI-driven predictive models can also forecast energy demand and adjust the electrolysis process accordingly, ensuring that hydrogen is produced when it is needed most.

2. **Advanced Materials Discovery** One of the challenges of electrolysis is finding the right materials for electrodes and catalysts that can withstand the high temperatures and voltages involved in the process. AI can play a crucial role in discovering new materials by simulating chemical reactions and predicting which materials will perform best in electrolysis. This could lead to the development of more efficient, durable, and cost-effective materials for hydrogen production.

3. **Scaling Up Electrolysis** As the demand for green hydrogen increases, there will be a need to scale up electrolysis systems. AI can help with this process by optimizing the design and operation of large-scale electrolyzers. By using machine learning to analyze data from existing systems, AI can identify patterns that will allow for more efficient scaling, reducing the time and cost associated with deploying larger electrolysis plants.

4. **Integration with Renewable Energy Sources** AI can also facilitate the integration of renewable energy sources with electrolysis systems. For example, AI can predict periods of high solar or wind energy generation and adjust the electrolysis process to take advantage of these times. This would help ensure that electrolysis systems operate at full capacity when renewable energy is abundant, increasing the overall efficiency of the hydrogen production process.

Conclusion

AI is rapidly becoming a transformative force in the energy sector, with applications spanning from energy production and storage to distribution and consumption. In the context of hydrogen energy production, AI has the potential to optimize the electrolysis process, making it more efficient and cost-effective. By leveraging AI's capabilities in real-time monitoring, predictive modeling, and materials discovery, we can unlock the full potential of green hydrogen and accelerate the transition to a sustainable, low-carbon energy future. The next frontier in AI-driven electrolysis is the integration of human-centric systems, which will be explored in the following chapters as we look toward a future where hydrogen energy is not only a clean solution for industries but also a personal source of power.

Chapter 5: The Future of Human Energy: Electrolysis and the Human Body

The concept of harnessing electrolysis directly in the human body is both fascinating and complex. At the intersection of cutting-edge technology, biology, and energy systems, this idea invites speculation and inquiry into how we might one day leverage human physiology as a source of energy. This chapter explores the hypothesis of humans being able to harness electrolysis directly, examines the biological systems that already perform energy conversion, and delves into the ethical and practical considerations of integrating AI with the human body for energy production.

Hypothesis: Can Humans Harness Electrolysis Directly?

When we think about energy production in humans, our first instinct might be to consider the body's inherent ability to produce energy through biochemical processes. For example, cells produce energy in the form of adenosine triphosphate (ATP) through cellular respiration, which involves the conversion of glucose and oxygen into energy. But what if, beyond biological processes, the human body could tap into electrolysis—the splitting of water molecules into hydrogen and oxygen—to produce additional energy?

This hypothesis opens a world of possibilities. Electrolysis in the body could theoretically function similarly to how the body produces energy in its cells, but using water as a medium for energy generation. If humans could somehow harness this energy directly, we might see a future where the body can supplement its natural energy sources with externally sourced hydrogen, unlocking new levels of performance, endurance, or self-sufficiency.

The key challenge here is whether the body can sustain the electrolysis process without disrupting vital systems. Electrolysis requires specific conditions, such as a significant voltage and the presence of an electrolyte to facilitate the flow of current. The question becomes whether the human body can support these conditions safely, and how AI-driven technologies could manage and optimize this process.

Biological Systems and Energy Conversion Mechanisms in the Human Body

Before we explore the feasibility of AI-driven electrolysis in humans, it's important to understand how the human body already converts and manages energy.

1. **Cellular Respiration**: The human body generates energy primarily through cellular respiration. This process takes place in the mitochondria, the powerhouse of the cell, where glucose and oxygen are combined to produce ATP. The process involves electron transfer chains that move electrons across membranes, releasing energy in the form of ATP. This process is, in many ways, analogous to electrolysis in that it involves electron flow, but on a biochemical level rather than an electrical one.

2. **ATP Production**: ATP is the main source of energy for cellular functions, from muscle movement to DNA replication. In essence, every biological activity requires ATP to function. If electrolysis could be integrated into the human system, it could theoretically serve as an additional means to generate ATP, allowing the body to sustain itself longer or provide energy for enhanced physical and cognitive performance.

3. **Electrochemical Gradients**: The human body also utilizes electrochemical gradients to manage energy. Nerve cells, for instance, create electrical signals through the movement of ions across membranes. Similarly, the body's cells generate power through electrochemical processes, such as the transport of sodium and potassium ions in and out of cells, which is essential for maintaining cellular function. While not the same as electrolysis, these processes highlight the body's existing capability to manage electrical charges and energy conversion.

The question remains: Can we augment or expand this system with AI and electrolysis-based technology to improve human energy production? Could AI control and optimize such a system, enabling the safe, efficient conversion of water into hydrogen for energy production?

Ethical and Practical Considerations for AI and Human Integration

Integrating AI-driven electrolysis into the human body raises several important ethical, practical, and societal questions that must be addressed. While the potential benefits could be transformative, it is essential to consider the implications for human health, safety, and autonomy.

1. **Safety and Biocompatibility**: The first and most important concern is the safety of integrating such technology into the human body. Electrolysis requires specific electrical conditions to function. Introducing such systems into human tissue must be done with extreme caution, ensuring that the processes do not cause damage to cells, tissues, or organs. Furthermore, the materials used in electrolysis systems must be biocompatible to avoid immune responses or toxic reactions.

2. **Energy Regulation and Control**: For AI to regulate and optimize electrolysis in the body, sophisticated systems would need to be developed to monitor and control energy production in real-time. AI models could ensure that the body's energy demands are met without overwhelming its natural processes. However, the risks of AI malfunctions or hacking would need to be minimized. Ensuring secure communication between the human body and external AI systems would be paramount to prevent misuse or errors.

3. **Autonomy and Privacy**: The idea of implanting AI-driven systems within the human body brings up significant concerns about autonomy and privacy. Who controls the technology that powers a person's energy production? Would individuals be able to turn off or adjust these systems if desired? Furthermore, the constant monitoring and data collection necessary to optimize AI-driven electrolysis could raise questions about personal privacy and consent.

4. **Long-Term Effects and Human Enhancement**: The integration of AI-driven electrolysis could be seen as a form of human enhancement. While the ability to harness additional energy might be appealing, it also raises ethical questions about the fairness of such enhancements. Who gets access to these technologies? Could they create a new form of inequality between those who can afford AI-powered energy systems and those who cannot?
5. **Health Implications**: There would also be long-term health considerations. What impact would AI-controlled electrolysis have on the human body's metabolism, immune system, or overall health? Would prolonged exposure to electrolysis-generated energy have unintended consequences? These questions must be thoroughly researched and understood before any integration takes place.

AI–Driven Electrolysis: A Bridge Between Technology and Biology

AI's potential to manage electrolysis processes within the human body is an exciting prospect. Through the use of AI models, the process could be optimized to ensure that energy production aligns with the body's natural rhythms and energy needs. AI could monitor cellular health, predict energy demands, and adjust the electrolysis process accordingly. Machine learning algorithms could also help predict the most efficient parameters for energy production, ensuring that the body's energy levels remain stable and sustainable.

Furthermore, AI-assisted bioengineering could bridge the gap between biology and technology. For example, AI could help develop bio-implants or micro-devices that work in harmony with the human body to enhance energy production. Such devices could utilize electrolysis in a localized area, such as the muscle tissue, to provide additional energy for physical exertion or recovery.

Another exciting possibility lies in the creation of hybrid systems that integrate biological processes with artificial technologies. For instance, AI could help design systems that mimic the natural processes of ATP production, while augmenting them with electrolysis to provide a supplementary energy source.

Conclusion

The possibility of integrating AI-driven electrolysis into the human body represents an exciting frontier in both human energy systems and biotechnology. While much work remains to be done to understand the feasibility, safety, and ethics of such systems, the potential benefits are immense. AI could help humans harness additional energy from water, potentially unlocking new levels of endurance, performance, and self-sufficiency.

However, as with any revolutionary technology, careful consideration of the ethical, practical, and health-related aspects is crucial. The future of AI-driven electrolysis in humans lies at the intersection of biology, technology, and human enhancement. As we move forward, ongoing research, regulation, and dialogue will be essential to ensure that this transformative technology is developed responsibly, for the benefit of all.

Chapter 6: AI's Role in Human–Centered Electrolysis Systems

The integration of artificial intelligence (AI) into human-centered electrolysis systems offers a unique intersection of advanced technology, biology, and energy production. AI is poised to optimize, enhance, and potentially revolutionize the way electrolysis can be used within the human body, transforming how energy is generated, managed, and utilized. This chapter explores the potential of AI to manage and enhance electrolysis processes, create predictive models for energy output, and bridge the gap between biology and technology through AI-assisted bioengineering.

Exploring AI's Potential in Managing and Enhancing Electrolysis Processes

At its core, AI has the capacity to process vast amounts of data and make real-time decisions based on complex, dynamic inputs. This ability makes AI uniquely suited to managing electrolysis systems, especially in scenarios that involve the human body. The complexity of the human body, with its fluctuating energy demands, biochemical reactions, and natural processes, requires an adaptive, intelligent system to optimize energy generation safely.

1. **Real-time Monitoring and Adjustment**: AI systems could continuously monitor the electrolysis process, adjusting variables such as voltage, current, and temperature in real time to ensure maximum efficiency. The body's energy needs could be tracked via sensors that detect biochemical signals or neural responses, providing a feedback loop that informs the AI when more or less energy is required. AI could ensure that electrolysis operates at optimal efficiency without disrupting normal bodily functions.

2. **Minimizing Energy Loss**: One of the challenges in electrolysis is minimizing energy loss, which is typically a concern with any energy conversion system. AI's ability to predict and adjust system parameters based on real-time data could ensure that the energy input is always aligned with the energy needs of the body. By optimizing power usage and reducing waste, AI could make electrolysis a viable, efficient energy source for the body.

3. **Personalized Energy Production**: Humans have unique energy requirements based on their activity levels, metabolic processes, and overall health. AI could tailor electrolysis-based energy production to meet these needs, ensuring that energy is generated efficiently during times of high activity or metabolic demand, and scaled back during rest periods. This personalized approach could allow the body to harness energy when needed without overstimulating natural processes.

Creating AI Models to Predict Energy Output and Optimize Efficiency

To harness the full potential of AI-driven electrolysis, the development of predictive models would be essential. These models would learn to understand and predict how the body responds to various energy demands and biochemical signals, allowing for better control over the electrolysis process. Machine learning algorithms would be key in this predictive process, enabling systems to evolve and adapt based on observed data over time.

1. **Data Collection and Learning**: Machine learning systems could begin by collecting data from various sensors within the body that track physiological markers such as heart rate, oxygen levels, and muscle activity. These markers could serve as proxies for energy demands. Over time, the AI would analyze patterns in this data, learning to predict when the body needs additional energy or when it can operate efficiently without relying on external energy production.

2. **Energy Forecasting**: Once AI models are developed, they could be used to forecast future energy requirements based on both short-term factors (such as exercise or stress) and long-term factors (such as health conditions or daily routines). The system could then adjust the electrolysis process to deliver energy exactly when and where it's needed, ensuring that the body remains optimally energized at all times.

3. **Optimization Algorithms**: The efficiency of the electrolysis system would depend on the ability of AI to optimize the interaction between the electrolysis process and the body's natural energy production systems. AI could use optimization algorithms to ensure that energy is produced without compromising the body's homeostasis. For instance, during physical exertion, the AI system could increase hydrogen production to support muscle activity, while during rest, it could lower output to conserve resources.

AI-Assisted Bioengineering: The Bridge Between Biology and Technology

AI-assisted bioengineering is the bridge that could seamlessly integrate electrolysis technology with the human body. By leveraging AI's computational power and bioengineering's innovative approach to human augmentation, it may become possible to create systems that enhance the human body's natural energy production without disrupting biological processes.

1. **Bio-Inspired Electrolysis Devices**: AI can help design electrolysis devices that mimic the natural processes of the human body. For example, AI could assist in designing micro-devices or implants that utilize electrolysis to generate hydrogen directly within specific tissues, such as muscles or the brain, without interfering with normal bodily functions. These devices could function like an artificial energy organ, supplementing the body's own energy systems.

2. **Biocompatible Materials and Microdevices**: AI could also drive the development of biocompatible materials used in bio-energy systems. These materials would need to be both functional in terms of conducting electrolysis reactions and non-toxic to the human body. AI's ability to simulate molecular interactions and predict material behaviors could drastically shorten development timelines for creating safe, effective bio-energy devices that are compatible with human tissue.

3. **Smart Implants and Neural Integration**: Smart implants powered by AI could take on the task of adjusting and monitoring electrolysis systems in the body. These implants would collect data on the body's energy needs, and AI would process this data to make adjustments. Neural interfaces could allow the AI to communicate with the brain, providing real-time feedback about energy availability and enhancing the body's natural energy regulation.

4. **Cellular-Level Energy Generation**: AI could also enable the creation of systems that work at the cellular level to optimize energy production. By designing bioengineered cells or artificial organelles that work in conjunction with the body's existing cellular processes, AI could help the body generate hydrogen energy directly from water molecules inside cells, potentially leading to highly efficient internal energy generation.

Ethical and Practical Considerations of AI in Human-Centered Electrolysis Systems

As with any cutting-edge technology that integrates AI into human biology, the integration of AI-driven electrolysis systems raises important ethical and practical concerns.

1. **Safety and Security**: The primary concern is ensuring the safety and security of AI-driven bio-energy systems. If AI systems are to manage and control electrolysis processes within the body, they must be secure from hacking or malfunction. The risks of system failure—ranging from energy imbalances to tissue damage—must be mitigated. Moreover, security measures must be in place to protect individuals from unauthorized access or control of their bio-energy systems.

2. **Privacy and Consent**: Since AI-driven electrolysis would require constant data collection from within the body, there are significant concerns around privacy. How would individuals' biological data be stored, used, and shared? Who controls the data generated by such systems, and how would consent be managed? These questions must be addressed to ensure that personal autonomy is respected.

3. **Bioethics**: The ethical implications of augmenting human bodies with AI-driven systems cannot be ignored. While AI has the potential to enhance human abilities and make life more efficient, it also raises questions about the limits of human enhancement. Should there be boundaries to how far AI is integrated into the body? How do we prevent the creation of inequality or the exploitation of such technology?

4. **Health and Long-Term Effects**: The long-term health effects of using electrolysis to supplement human energy are still largely speculative. AI's role in managing these systems would be essential in monitoring their impact on overall health and ensuring that no adverse long-term effects occur. Ongoing research into the physiological and biochemical impacts of these technologies is necessary to fully understand the risks involved.

Conclusion

AI has the potential to revolutionize human-centered electrolysis by optimizing the production and management of energy within the body. Through real-time monitoring, predictive modeling, and AI-assisted bioengineering, we can imagine a future where humans are able to harness additional energy through electrolysis. However, integrating such technologies requires careful consideration of safety, ethics, and long-term health effects. As AI continues to evolve and intersect with bioengineering, the possibilities for enhancing human energy systems and capabilities are immense, but the path forward must be navigated responsibly and thoughtfully.

Chapter 7: The Physics and Chemistry of Electrolysis: Advanced Concepts

To fully understand the potential of AI-driven electrolysis in human energy systems, it is essential to grasp the advanced principles of electrolysis from both a physics and chemistry perspective. Electrolysis is not just about splitting water molecules—it involves complex processes that depend on the interaction of electrical energy, chemical reactions, and materials. This chapter delves into the molecular level of electrolysis, the factors affecting its efficiency, and the innovations that could make electrolysis more viable for human integration.

The Molecular Level of Electrolysis

At its core, electrolysis involves the breaking of bonds between atoms in a molecule using electrical energy. When applied to water (H_2O), the process splits the molecule into its constituent elements: hydrogen (H_2) and oxygen (O_2). This occurs in two half-reactions at the cathode and anode of the electrolytic cell.

1. **Cathode (Reduction Reaction)**: At the cathode, the process of reduction occurs. Water molecules are reduced by gaining electrons (reduction), resulting in the formation of hydrogen gas. The reaction can be expressed as:

   $$2H_2O + 2e^- \rightarrow H_2 + 2OH^-$$

   In this reaction, two water molecules (H_2O) gain two electrons to form hydrogen gas (H_2) and hydroxide ions (OH^-).

2. **Anode (Oxidation Reaction)**: At the anode, water molecules are oxidized by losing electrons (oxidation), producing oxygen gas and releasing protons (H^+) into the solution. This reaction can be expressed as:

   $$2H_2O \rightarrow O_2 + 4H^+ + 4e^-$$

   In this reaction, two water molecules lose four electrons to produce oxygen gas (O_2) and protons (H^+).

Overall, the combined electrolysis of water leads to the following reaction:

$$2H_2O \rightarrow 2H_2 + O_2$$

This reaction, while conceptually simple, involves significant energy changes and the movement of electrons, protons, and ions. The energy required to drive these reactions is supplied by an external power source, typically electricity.

Key Factors Affecting Efficiency

The efficiency of electrolysis is influenced by several factors, which must be optimized to make the process suitable for use in human-centered energy systems. These factors include the voltage, temperature, electrolyte composition, electrode materials, and system design.

1. **Voltage and Current**: The voltage applied across the electrolyzer determines how effectively water is split into hydrogen and oxygen. To initiate the electrolysis of water, a minimum voltage—typically around 1.23 V—is required. However, to overcome various resistances in the system, higher voltages are often used, which increases the overall energy consumption. The amount of current supplied also impacts the rate of hydrogen production. Too much current can lead to inefficiencies or overconsumption of power, while too little will slow down the process.

2. **Electrolyte Composition**: Water by itself is a poor conductor of electricity. Therefore, an electrolyte (usually a solution of salts, acids, or bases) is added to increase the conductivity of water and facilitate the flow of ions. The choice of electrolyte affects the efficiency of the electrolysis process. For example, alkaline electrolytes (such as potassium hydroxide) are often used in industrial electrolysis systems because they provide high conductivity and efficiency. However, the electrolyte's composition can also influence the rate of corrosion of the electrodes, which is an important factor in the long-term viability of electrolysis systems.

3. **Electrode Materials**: The electrodes used in the electrolysis process must be conductive and resistant to corrosion, especially since the process generates highly reactive gases. Materials such as platinum, iridium, or nickel are often used for the electrodes, but these can be expensive. Researchers are exploring more cost-effective and durable materials, including carbon-based electrodes or novel nanomaterials, to reduce the overall cost and improve the performance of electrolyzers.

4. **Temperature**: Electrolysis is an exothermic process, meaning it generates heat. Operating electrolysis at higher temperatures can increase the rate of the reaction by providing more energy to overcome activation barriers. However, higher temperatures can also increase the risk of system degradation and reduce the lifespan of electrodes and electrolytes. Balancing temperature with efficiency and longevity is crucial in optimizing electrolysis for human systems.

5. **System Design**: The design of the electrolyzer itself—whether it is a small, portable device or a large industrial system—also plays a significant role in efficiency. Innovations in membrane technology, cell configuration, and energy recovery systems can all contribute to the performance of electrolysis systems. For human integration, miniaturized systems that are highly efficient and capable of operating within the body's constraints would be essential.

Innovations in Electrolysis Technologies

To make electrolysis more practical for human-centered energy systems, innovations in electrolysis technology are necessary. These innovations focus on improving efficiency, reducing costs, and enabling smaller, more adaptable systems. Some key advancements include:

1. **Proton Exchange Membrane (PEM) Electrolysis**: Proton exchange membrane (PEM) electrolysis uses a solid polymer electrolyte instead of a liquid electrolyte, which increases efficiency and allows for faster response times. PEM electrolysis is particularly well-suited for use with renewable energy sources, as it can operate efficiently under varying electrical input. This technology could be critical in developing small-scale, portable electrolysis devices for human integration, where space, weight, and efficiency are paramount.
2. **High-Temperature Electrolysis**: High-temperature electrolysis (HTE) operates at temperatures above 700°C and offers the potential for higher efficiencies due to the reduced electrical energy required at elevated temperatures. HTE can be integrated with industrial processes or even waste heat sources, making it a potential solution for both large-scale and human-centered energy systems. Research is ongoing into materials that can withstand the high temperatures required for HTE without degrading.

3. **Electrolysis Using Renewable Energy**: Electrolysis can be powered by renewable energy sources such as solar, wind, or hydroelectric power, creating a closed-loop, sustainable energy cycle. This is particularly important in the context of AI-driven electrolysis for humans, as the energy used to power electrolysis would ideally come from renewable sources. AI could optimize the integration of renewable energy sources with electrolysis systems, ensuring that hydrogen production occurs during times of abundant energy availability and at optimal efficiency.

4. **Nanotechnology and Catalysts**: The use of nanomaterials and advanced catalysts is revolutionizing electrolysis technology. Researchers are exploring how to enhance the efficiency of electrolysis through the use of nanostructured electrodes that provide a larger surface area for reactions to occur. Catalysts, such as nickel or cobalt-based compounds, are being developed to reduce the energy required for water splitting, making electrolysis more energy-efficient and cost-effective.

Scalability and Efficiency in Human-Centered Electrolysis

In the context of human energy systems, scalability and efficiency are critical factors. While large-scale industrial electrolysis systems are already in use for hydrogen production, the goal is to adapt these technologies for human-centered applications, where size, weight, and power requirements are more constrained.

1. **Miniaturization**: For electrolysis to be useful within the human body, it must be miniaturized while maintaining efficiency. This requires advancements in membrane and electrode technology, as well as the development of lightweight, biocompatible materials that can withstand the harsh environment of the human body.

2. **Energy Balance**: The energy generated through electrolysis must be carefully balanced with the energy required to operate the system. AI can help optimize this balance by predicting when energy is needed and adjusting electrolysis processes accordingly. For example, during periods of high physical activity, AI could increase hydrogen production to fuel muscles, while reducing output during periods of rest to conserve resources.

3. **Integration with Biological Systems**: Electrolysis systems for human integration would need to be highly adaptable, able to respond dynamically to the body's changing energy demands. AI-driven systems could optimize energy production in response to real-time physiological feedback, ensuring that the body's natural energy systems are not overwhelmed or disrupted.

Conclusion

The advanced physics and chemistry of electrolysis provide the foundation for AI-driven systems that can transform the way energy is generated and utilized within the human body. By optimizing key factors such as voltage, temperature, and electrolyte composition, and by incorporating innovations in technology, electrolysis can become a viable method for supplementing human energy production. As AI continues to advance, it will play a crucial role in managing these complex processes, ensuring that electrolysis systems operate efficiently and safely within the human body. The future of AI-driven electrolysis lies in the continuous refinement of these technologies, making the dream of sustainable, human-centered energy production a reality.

Chapter 8: AI-Driven Electrolysis in Medical Technology

AI-driven electrolysis holds incredible potential not only in the energy sector but also in the realm of healthcare and medicine. By harnessing the capabilities of electrolysis in combination with artificial intelligence, we could revolutionize medical treatments, power medical devices, and even introduce new possibilities in disease treatment and prevention. This chapter delves into the transformative ways AI-driven electrolysis could impact healthcare, including the development of implantable devices, personalized medicine, and the intersection of energy solutions and medical advancements.

How AI-Driven Electrolysis Could Impact Healthcare and Medicine

The medical field is continuously seeking new technologies that can optimize patient care, enhance the effectiveness of treatments, and increase accessibility to vital services. AI-driven electrolysis could play a pivotal role in several medical areas, including the development of self-sustaining medical devices, energy-efficient healthcare technologies, and even new forms of therapy.

1. **Self-Powered Medical Devices**: Many modern medical devices, such as pacemakers, deep brain stimulators, and insulin pumps, require a reliable and constant power supply to function. Traditional power sources, such as batteries, need to be replaced periodically, which can be costly and invasive. By integrating AI-driven electrolysis systems, medical devices could potentially generate their own energy directly from the body.

   Electrolysis, powered by bioelectricity, could convert water in the body into hydrogen, providing a sustainable energy source to power medical implants and devices. AI could optimize the electrolysis process, adjusting energy production according to the device's needs, ensuring constant, autonomous operation without the need for battery replacements.

   Imagine a pacemaker that not only monitors heart rhythm but also draws energy from the body's natural processes to keep functioning, eliminating the need for invasive surgeries to replace batteries. AI could help regulate the amount of energy needed for such devices, ensuring optimal performance while minimizing energy wastage.

2. **Personalized Medicine and Energy Production**: AI-driven electrolysis could also play a role in personalized medicine by tailoring energy production to the unique metabolic needs of individuals. By analyzing a patient's physiological data, AI systems could predict energy needs based on real-time variables such as age, health status, activity levels, and disease progression.

For example, in cases where a patient has a metabolic disorder or a condition that affects energy levels, AI could adjust electrolysis systems to provide supplementary energy to maintain body functions. This could help alleviate symptoms of chronic fatigue, improve recovery times for patients undergoing surgeries, or even enhance the body's ability to fight infections.

In personalized healthcare, AI-driven electrolysis could be used as part of a treatment plan to address energy deficiencies, especially for conditions such as mitochondrial disorders, which impair the body's ability to generate ATP. By harnessing electrolysis, the body could generate its own energy in a more efficient and sustainable manner.

3. **Wound Healing and Tissue Regeneration**: AI-powered electrolysis could potentially be used in medical devices that stimulate tissue regeneration and wound healing. Electrolysis creates charged particles (ions) and molecules (such as hydrogen and oxygen), which can have therapeutic effects on cells and tissues.

For instance, when used in wound care, electrolysis could be used to generate a microenvironment of reactive oxygen species (ROS) at the site of injury. ROS are known to promote cellular repair and regeneration, aiding the healing process. By controlling the concentration and timing of these reactive molecules, AI systems could optimize the wound healing process, providing faster recovery and reducing the risk of infection.

Similarly, AI-driven electrolysis could enhance the regeneration of tissues in patients with degenerative diseases, such as osteoarthritis or spinal cord injuries. By promoting cellular growth and stimulating the release of growth factors, electrolysis could support the body's natural healing capabilities and assist in tissue regeneration.

4. **Advanced Drug Delivery Systems**: AI-driven electrolysis could also play a critical role in the development of advanced drug delivery systems. These systems could use electrolysis to release drugs or therapeutic agents in a controlled manner at the site of action, minimizing side effects and improving the precision of treatments.

For example, electrolysis could power nanopumps that transport medications through the bloodstream or directly to specific tissues. AI could manage the timing, dosage, and location of the drug release, ensuring that the therapeutic agent is delivered precisely where it's needed at the right time. This could be particularly beneficial for cancer treatments, where precise targeting of tumors can reduce damage to surrounding healthy tissue.

5. **Electrolysis-Based Biofeedback Systems**: In addition to its use in medical devices, electrolysis could be integrated into biofeedback systems that allow patients to monitor and optimize their health in real-time. For example, patients with chronic conditions such as hypertension, diabetes, or cardiovascular disease could wear devices that use electrolysis to generate small amounts of energy for internal monitoring and self-regulation.

AI-powered biofeedback systems could track a patient's physiological responses, such as blood pressure, glucose levels, or heart rate, and provide real-time guidance for making lifestyle or medication adjustments. By incorporating electrolysis, these systems could operate autonomously, ensuring continuous monitoring without the need for external power sources.

6. **Impact on Mental Health and Cognitive Function**: Recent research has shown that the brain's electrical activity plays a significant role in mood regulation, cognition, and overall mental health. AI-driven electrolysis could be used in therapeutic applications to modulate brain activity, such as in the treatment of depression, anxiety, or even neurodegenerative diseases like Alzheimer's.

   By generating hydrogen and oxygen molecules in a controlled manner, electrolysis could help stimulate areas of the brain associated with mood regulation and cognitive function. AI could adjust the electrolysis process to provide specific energy inputs to the brain, potentially improving neural connectivity, stimulating neurogenesis, and promoting mental clarity.

   Electrolysis could also be used in conjunction with brain-computer interfaces (BCIs) to directly stimulate brain activity. For instance, AI could monitor the brain's electrical patterns and modulate the energy production within the brain to improve cognitive performance or alleviate symptoms of mental health disorders.

The Intersection of Energy Solutions and Medical Advancements

As AI-driven electrolysis continues to evolve, the intersection between energy solutions and medical advancements will become more pronounced. AI's ability to optimize energy production for medical applications opens up possibilities for a wide range of healthcare innovations.

1. **Portable and Wearable Medical Devices**: One of the most promising applications of AI-driven electrolysis is the development of portable, wearable medical devices that operate independently of external power sources. These devices could include smart insulin pumps, heart monitors, or even wearable health trackers that generate their own power, eliminating the need for constant charging or battery replacements.

2. **Remote Monitoring and Telemedicine**: AI-driven electrolysis could enable remote monitoring of patients in home settings, particularly for those with chronic illnesses or those requiring continuous monitoring, such as cardiac patients or individuals with diabetes. By combining electrolysis with AI-powered data analysis, patients could receive continuous care without the need for frequent hospital visits, empowering both patients and healthcare providers.

3. **Sustainability in Healthcare**: Integrating electrolysis into medical technologies could also make healthcare more sustainable. The use of self-sustaining medical devices powered by the body's natural processes would reduce reliance on non-renewable resources, such as disposable batteries. Moreover, AI could help optimize the use of renewable energy in medical facilities, reducing hospitals' carbon footprints and helping them become more environmentally responsible.

Conclusion

AI-driven electrolysis presents a unique and transformative opportunity to improve healthcare delivery, enhance patient care, and create self-sustaining medical systems. From implantable devices that generate their own power to personalized treatment plans that adjust energy production based on real-time needs, the potential applications in medicine are vast and diverse. As AI and electrolysis technology continue to advance, we can expect a future where the two fields converge to create more efficient, effective, and sustainable healthcare solutions for patients around the world. However, as with any new technology, the integration of AI-driven electrolysis into medical systems must be approached with caution, ensuring patient safety, regulatory compliance, and ethical considerations are met.

Chapter 9: Designing Bio-Energy Systems for Human Integration

The integration of AI-driven electrolysis into human systems requires the creation of bio-energy systems that are both efficient and compatible with the human body. Designing these systems is a complex process that involves blending advanced technologies with biological functions, requiring interdisciplinary knowledge across bioengineering, nanotechnology, AI, and medicine. This chapter explores the essential components of bio-energy systems, challenges in bioengineering and tissue compatibility, and the potential for exploring energy generation at the cellular level.

Building Sustainable Systems That Combine Human Biology with AI-Powered Electrolysis

A successful bio-energy system for human integration hinges on the seamless fusion of biological and technological components. The primary objective is to design systems that complement the human body's existing energy processes rather than disrupt them. This involves creating bio-energy devices that harness the power of AI-driven electrolysis to generate energy autonomously, utilizing water and the body's natural bioelectric processes.

1. **Autonomous Energy Production**: The key innovation in designing bio-energy systems is developing self-sustaining energy devices that do not rely on external power sources. These devices would utilize electrolysis to produce hydrogen and oxygen from water within the human body. Hydrogen can then be converted into energy through fuel cells or other mechanisms to power medical implants, bio-enhanced devices, or even assist in human bodily functions that require extra energy.

   AI can optimize the energy generation process in real time, ensuring that the body receives the precise amount of energy needed for various activities. This self-sufficiency could eliminate the need for external power sources, such as batteries or frequent maintenance, making the system more reliable and efficient.

2. **Biocompatible Electrolysis Systems**: The materials used in these systems must be biocompatible, ensuring they do not cause an immune response or damage body tissues. The development of these materials requires research into substances that can perform electrolysis in the body without toxicity. AI can help in the identification and design of such materials by simulating molecular interactions and testing various compounds for their biocompatibility and efficiency. Furthermore, AI-driven systems can predict how these materials will interact with body fluids and cells, enabling the creation of bio-energy devices that can integrate into the body seamlessly without adverse effects.

3. **Wearable Bio-Energy Devices**: In addition to implantable devices, AI-powered bio-energy systems can be developed as wearable technologies. These devices could harness water from the skin's surface or sweat, facilitating small-scale energy generation for personal use. For example, a wearable device like a smart band or patch could integrate electrolysis cells to power wearable sensors or other low-energy devices, enabling continuous, autonomous energy generation. Wearable bio-energy devices could also be designed to monitor and respond to the body's energy requirements in real time, using AI to optimize power distribution for physical or mental tasks. These devices would offer a portable solution for humans who require enhanced energy management or additional energy support during periods of high stress or physical exertion.

Challenges in Bioengineering and Tissue Compatibility

The primary challenge in integrating AI-driven electrolysis into the human body lies in ensuring the system's compatibility with human tissues. Biological systems have evolved to function within specific conditions, and any technology introduced must work in harmony with these existing processes.

1. **Tissue Integration**: The introduction of electrolysis technology within the human body must be done in a way that does not disrupt tissue function or compromise the body's integrity. Electrolysis systems must be carefully designed to work in harmony with the body's natural energy processes, such as cellular respiration and ion transport. For example, the process of electrolysis generates small amounts of heat and gas (hydrogen and oxygen), which could have negative effects on the surrounding tissues if not properly managed.

   AI can play a crucial role in regulating the heat and gas produced during electrolysis to ensure that it is absorbed by the body in a controlled manner. Advanced AI models could simulate how these factors interact with biological systems, allowing engineers to design systems that minimize risks while maximizing efficiency.

2. **Minimizing Invasive Procedures**: One of the critical concerns when integrating technology into the human body is minimizing the invasiveness of procedures. Implanting devices that require regular maintenance or involve complex surgeries can lead to health risks such as infections, tissue rejection, and complications in recovery.

AI-driven electrolysis systems could be designed to reduce the invasiveness of integration. For example, bio-energy systems could be implanted using minimally invasive techniques, relying on AI to manage system calibration and performance remotely. Wearable or even non-invasive devices, such as patches or flexible electronics, could also provide a less intrusive solution while still allowing for energy generation and bio-monitoring.

3. **Long-Term Biocompatibility and Longevity**: The long-term effects of implanting AI-driven electrolysis systems into the body must be carefully studied. As with any implantable technology, the materials and devices used must remain effective over time without causing adverse effects such as tissue damage, inflammation, or immune responses. AI can facilitate this process by analyzing and predicting potential issues related to long-term implantation and monitoring for signs of wear and tear on devices.

   Furthermore, AI-powered systems could adapt over time, learning from ongoing data to predict the maintenance needs of bio-energy systems. This would reduce the risk of unexpected malfunctions and extend the lifespan of the device, ensuring its long-term functionality.

Exploring Energy Generation at the Cellular Level

The ultimate goal of AI-driven electrolysis in human systems may be to bring energy production to the cellular level. Just as the mitochondria within cells generate energy through oxidative phosphorylation, AI could be used to enhance cellular energy generation by incorporating electrolysis mechanisms.

1. **Cellular Bioenergetics**: By using AI to monitor and manage cellular bioenergetics, it could be possible to create micro-devices that enhance or mimic natural cellular energy production through electrolysis. For example, bio-energy systems could be designed to create hydrogen at the cellular level, directly fueling cellular functions. This would allow energy production to be closely aligned with the body's metabolic needs, optimizing energy usage in specific tissues, such as muscle cells during exercise or neural cells during periods of intense cognitive activity.

2. **Nanotechnology and Bio-Energy**: Nanotechnology could play a significant role in bio-energy systems by enabling the creation of nano-scale electrolysis devices that work inside the human body. These devices could be designed to generate energy within individual cells or tissues, powered by AI to manage the efficiency of energy production.

By integrating AI with nanotechnology, it is possible to design self-regulating systems that adjust energy production based on real-time data from the body. For example, when a muscle cell requires additional energy during physical activity, AI could prompt the nano-scale electrolysis device to generate more hydrogen, which would be converted into usable energy for the cell.

Conclusion

Designing bio-energy systems that integrate AI-driven electrolysis into human biology is an exciting frontier in both bioengineering and energy production. While challenges such as tissue compatibility, device longevity, and minimizing invasiveness must be addressed, the potential for creating autonomous, self-sustaining energy systems within the body is vast. Through innovative AI models, biocompatible materials, and advanced bioengineering techniques, AI-powered electrolysis systems could one day provide humans with enhanced energy capabilities, reducing dependence on external power sources and enabling more efficient energy use. The convergence of AI, bioengineering, and energy production could redefine how we approach human health, performance, and energy management, opening up new possibilities for both medicine and energy solutions.

Chapter 10: Building the Future: From Concept to Prototype

Turning the concept of AI-driven electrolysis systems for human energy integration into a functioning prototype is one of the most crucial steps in realizing the potential of this transformative technology. While the theoretical underpinnings are in place, developing tangible systems requires innovation, research, and collaboration across various disciplines, including bioengineering, nanotechnology, and artificial intelligence. This chapter explores how AI-driven electrolysis systems could be prototyped and tested, the role of innovation in bringing these concepts to life, and the key milestones needed to translate theory into real-world applications.

How AI-Driven Electrolysis Systems Could Be Prototyped and Tested

The prototyping and testing of AI-driven electrolysis systems within human contexts involves a series of rigorous steps aimed at validating the technology's feasibility, safety, and efficiency. From early-stage designs to practical implementations, each step builds upon the previous one to ensure that the final product is both effective and secure.

1. **Design and Simulation**: The first step in prototyping AI-driven electrolysis systems is designing a system that combines bioengineering and AI optimization. This involves integrating electrolysis technologies with biological systems in a way that is efficient, biocompatible, and capable of generating energy within the human body.

Advanced AI models can be used to simulate various aspects of the electrolysis process, from material selection to system integration. By testing different configurations in a virtual environment, researchers can predict how the system will behave inside the body, optimizing parameters like power output, energy conversion efficiency, and waste management.

These simulations could also account for the dynamic nature of human biology. For example, AI could predict how the electrolysis system responds to various physiological conditions, such as changes in blood flow, temperature, or metabolic demand, ensuring that the system is adaptive and able to operate under different conditions.

2. **Prototype Development**: Once the system design has been validated through simulation, the next step is creating physical prototypes. In this phase, bio-engineered devices are built to incorporate the electrolysis technology and AI-driven monitoring systems. These prototypes can be small-scale models that demonstrate the core functionality of the system, such as hydrogen production, energy storage, and energy transfer.

The prototyping process will involve selecting appropriate materials that are biocompatible and capable of performing electrolysis without causing harm to the body. At the same time, AI systems embedded within the prototype will manage the electrolysis process, adjusting the energy output as necessary based on the surrounding physiological data.

Testing prototypes in controlled environments, such as lab-based simulations or animal models, will provide insights into how well the systems function. AI algorithms can adjust parameters in real-time, optimizing efficiency and safety based on the data collected during tests.

3. **Iterative Testing and Optimization**: After initial prototypes are developed, they will undergo a series of tests to evaluate performance in real-world scenarios. For example, prototypes might be tested in environments that simulate human biological conditions, such as varying temperatures, pressures, and fluid compositions. AI can continually refine the electrolysis process during these tests, learning from each iteration to improve the efficiency and safety of the system. Safety is a critical concern in this phase, as any malfunction in the electrolysis system could have unintended consequences for human health. AI can monitor the system in real-time, identifying potential issues such as overheating, excessive gas production, or electrochemical imbalances, and make adjustments as needed to prevent harm.

4. **Human Trials and Ethical Considerations**: Once a prototype passes preclinical testing, it can be considered for human trials. This stage requires rigorous ethical review and regulatory approval to ensure that the system does not pose any risks to human health. AI-powered electrolysis systems will need to meet high safety standards before being integrated into the human body.

   Human trials would begin with smaller, controlled groups to closely monitor how the technology integrates with biological systems. AI will play a key role in monitoring these trials, collecting data on user health and performance, and adjusting the systems for optimal results. Ethical considerations, including informed consent and the protection of patient privacy, will be central to this phase of development.

The Role of Innovation and Research in Developing Functional Systems

Innovation and research are key drivers in advancing AI-driven electrolysis systems from the drawing board to practical, functional devices. Several areas of research will be critical for achieving breakthroughs and improving the scalability of these systems.

1. **Interdisciplinary Collaboration**: Creating AI-driven electrolysis systems that are safe, efficient, and effective requires collaboration across multiple disciplines, including bioengineering, artificial intelligence, materials science, and medical technology. Researchers from diverse fields will need to work together to ensure that the technology can be integrated into the human body in a way that is compatible with existing biological systems.

For example, bioengineers will work to create biocompatible materials for electrolysis cells, while AI experts develop algorithms that can dynamically control the energy production process. Medical professionals will ensure that the technology is safe and effective for human use. This interdisciplinary collaboration will be essential to overcoming the numerous challenges associated with the technology's development.

2. **Material Science and Nanotechnology**: Materials science and nanotechnology will play a crucial role in the development of bio-energy systems that utilize electrolysis. Advances in nanomaterials may allow for the creation of micro-scale electrolysis devices that can operate efficiently in the body without causing harm to cells or tissues. Nanotechnology could also enable the creation of ultra-efficient electrolysis catalysts, reducing the energy required for hydrogen production and improving overall system performance.

   AI can be used to model and simulate the interactions between different nanomaterials, predicting which combinations will produce the best results. By combining the power of AI with cutting-edge material science, researchers can accelerate the development of next-generation electrolysis systems.

3. **Energy Storage and Conversion Efficiency**: Another critical area of research is energy storage. Once hydrogen is produced through electrolysis, it must be stored and converted into usable energy for the body. Researchers are exploring various methods of hydrogen storage, such as metal hydrides or carbon nanotube-based systems, that could be used in bio-energy applications.

AI can help optimize the energy conversion process, ensuring that the stored hydrogen is used efficiently. It can also dynamically manage the flow of energy within the system, directing it where it is most needed, such as to medical devices or muscle tissues during physical exertion. By optimizing both energy production and storage, AI can make electrolysis-based bio-energy systems more viable for long-term use.

Key Milestones Needed to Turn Theory into Real-World Applications

The journey from theoretical concept to real-world application involves achieving several key milestones. These milestones provide clear benchmarks for success, guiding the development of AI-driven electrolysis systems.

1. **Prototype Validation**: The first milestone is the successful development of a functional prototype that demonstrates the viability of AI-powered electrolysis within human systems. This prototype must undergo rigorous testing to ensure that it can operate safely and efficiently in a biological environment.
2. **Clinical Trials**: Once the prototype has been tested in lab and animal models, the next milestone is human clinical trials. Successful trials will demonstrate the technology's effectiveness and safety in humans, paving the way for regulatory approval.
3. **Commercialization and Widespread Adoption**: The final milestone is the commercialization of AI-driven electrolysis systems, making them available for medical, consumer, and industrial applications. This will require substantial investment in infrastructure, research, and production to ensure that the systems can be scaled up for widespread use.

Conclusion

Turning the vision of AI-driven electrolysis systems into reality is an ambitious and exciting goal that requires innovation, research, and collaboration. By prototyping and testing these systems, researchers can create devices that integrate seamlessly into the human body, offering sustainable, efficient energy solutions for a wide range of applications. As AI continues to evolve, it will play a pivotal role in optimizing electrolysis systems, paving the way for a future where humans can harness hydrogenic energy directly from water for medical, personal, and industrial purposes. The path from concept to prototype may be challenging, but the rewards of creating a self-sustaining, energy-efficient system are immense, with the potential to revolutionize how we generate and use energy in the future.

Chapter 11: AI and the Energy Crisis: Hydrogen as the Solution

The world is at a crossroads, facing a growing energy crisis that is inextricably linked to the global challenges of climate change, energy inequality, and the unsustainable nature of fossil fuel reliance. As the demand for clean and renewable energy sources intensifies, innovative solutions are more crucial than ever. Hydrogen, particularly when produced through AI-driven electrolysis, represents one of the most promising answers to this crisis. This chapter examines the global energy challenges we face, explores how hydrogen energy powered by AI can address these challenges, and discusses the long-term benefits of AI-driven hydrogen production systems in shaping a sustainable and equitable energy future.

Global Energy Challenges and the Importance of Finding Sustainable Alternatives

The global energy landscape is under immense pressure. Traditional fossil fuel sources, such as coal, oil, and natural gas, have been the primary drivers of economic growth and industrial development for centuries. However, these energy sources come at a significant environmental cost, contributing to climate change, air pollution, and environmental degradation. The negative impacts of these energy sources are most acutely felt by vulnerable populations in developing nations, but they also affect the entire planet's ecosystems and future generations.

In addition to environmental concerns, energy inequality remains a pressing issue. Approximately 800 million people worldwide still lack access to reliable electricity, primarily in sub-Saharan Africa and parts of South Asia. Addressing this energy gap is essential for achieving global economic and social development, but it must be done in a way that is both sustainable and equitable.

Moreover, the geopolitical instability associated with the global energy supply—often centered around fossil fuel extraction and distribution—adds another layer of urgency to finding alternative energy solutions. The reliance on oil and gas, for example, has fueled conflicts and strained international relations. A transition to clean, locally sourced energy could help mitigate these risks and promote global peace and stability.

As the world shifts away from fossil fuels, hydrogen stands out as a key alternative. Hydrogen is the most abundant element in the universe, and when produced sustainably, it offers a clean, versatile, and energy-dense solution that could address the myriad challenges facing the global energy system.

How Hydrogen Energy and AI Can Address These Challenges

Hydrogen, produced through electrolysis powered by renewable energy sources, presents an exciting solution to the energy crisis. Unlike fossil fuels, hydrogen can be produced with minimal environmental impact, especially when the electrolysis process is powered by clean, renewable energy sources such as wind, solar, or hydroelectric power. When hydrogen is used as a fuel, the only byproduct is water, making it a truly clean energy carrier.

AI plays a pivotal role in optimizing hydrogen production through electrolysis. Electrolysis, as a process, is energy-intensive, and its efficiency can be significantly improved with the help of AI. Here's how:

1. **Optimizing Electrolysis Efficiency**: AI algorithms can analyze data from various sensors and inputs to optimize the electrolysis process. By continuously adjusting parameters like voltage, temperature, and pressure, AI can ensure that electrolysis systems operate at peak efficiency, minimizing energy consumption while maximizing hydrogen production. Over time, AI systems can learn from the data and predict the optimal conditions for different environmental and operational factors.
2. **Smart Energy Management**: AI can also manage the integration of renewable energy sources (such as solar or wind) with hydrogen production systems. Since renewable energy generation is intermittent, AI can predict when renewable energy generation will peak or dip, allowing hydrogen production to be ramped up during times of excess energy and reduced during times of scarcity. This not only ensures the efficient use of renewable energy but also enables the storage of hydrogen as a clean energy source for later use, helping to stabilize the energy grid.

3. **Decentralized Hydrogen Production**: One of the key challenges in global energy distribution is the inefficiency and expense of transporting energy over long distances. AI-driven electrolysis systems could enable decentralized hydrogen production, where hydrogen is generated locally from water and renewable energy sources. This could eliminate the need for extensive energy transport infrastructure, reduce energy losses, and promote energy independence in remote or underserved regions. By using AI to manage and optimize decentralized systems, hydrogen production can be more easily scaled and tailored to local energy needs.

Long-Term Benefits of AI-Driven Hydrogen Production Systems

AI-driven hydrogen production systems offer numerous long-term benefits that can address both the immediate energy crisis and broader sustainability goals. These systems are designed to be flexible, scalable, and sustainable, making them an attractive option for transitioning to a clean energy future.

1. **Carbon Neutrality and Decarbonization**: One of the most significant benefits of hydrogen energy is its potential to contribute to global decarbonization efforts. Hydrogen produced through electrolysis using renewable energy sources produces no carbon emissions, making it a critical component in the transition to a net-zero carbon economy. AI can enhance the efficiency of hydrogen production systems, helping to further reduce carbon footprints and accelerate the global shift away from fossil fuels.

2. **Energy Storage and Grid Stabilization**: Hydrogen can be stored and transported relatively easily compared to electricity, making it an ideal candidate for long-term energy storage. AI-driven electrolysis systems can produce hydrogen during periods of low demand or excess renewable energy generation and store it for later use. This hydrogen can be converted back into electricity or used as a fuel for transportation, industry, or residential heating, helping to stabilize energy grids and provide reliable energy when renewables are not available.

3. **Economic Opportunities and Job Creation**: The hydrogen economy has the potential to create millions of new jobs in the sectors of energy production, storage, transportation, and technology development. By investing in AI-driven electrolysis systems, governments and industries can promote economic growth, support innovation, and create opportunities for skilled workers. The development of hydrogen infrastructure—such as refueling stations, storage facilities, and production plants—will also contribute to economic development, particularly in regions with access to renewable energy sources.

4. **Energy Security and Geopolitical Stability**: A global shift to hydrogen energy could help reduce dependence on fossil fuel imports, promoting energy security and geopolitical stability. By producing hydrogen locally from renewable sources, countries can reduce their reliance on energy imports, mitigating the risks associated with global energy supply chains and conflicts over fossil fuel resources. AI-driven systems can help optimize hydrogen production and distribution, ensuring that hydrogen energy is available to all regions regardless of their geographic location.

The Role of Policy, Investment, and Infrastructure in Scaling Hydrogen Solutions

To unlock the full potential of hydrogen as a clean energy solution, robust policy frameworks, strategic investments, and supportive infrastructure must be put in place. Governments, industries, and international organizations must work together to create a conducive environment for the widespread adoption of AI-driven hydrogen technologies.

1. **Policy and Regulation**: Governments play a crucial role in shaping the future of hydrogen energy through policy and regulation. Clear, long-term policies that incentivize investment in hydrogen infrastructure and research can accelerate the development and deployment of AI-driven electrolysis systems. Policies should also address the necessary safety standards, environmental regulations, and public health considerations to ensure the safe and efficient integration of hydrogen energy into national energy systems.

2. **Investment in R&D**: Continued investment in research and development is necessary to advance the technology behind AI-driven electrolysis systems. Funding for innovation in AI, hydrogen production technologies, storage methods, and distribution networks will help lower costs, improve efficiency, and accelerate commercialization. Collaboration between public and private sectors, as well as international cooperation, will be key to driving innovation and scaling hydrogen solutions globally.

3. **Infrastructure Development**: The successful implementation of hydrogen energy systems requires the development of a global hydrogen infrastructure, including production facilities, storage units, transportation networks, and distribution systems. AI can be used to optimize the design and operation of this infrastructure, ensuring that it is efficient, reliable, and scalable. Governments and industries must invest in creating this infrastructure to enable widespread access to hydrogen energy.

Conclusion

The energy crisis is one of the defining challenges of our time, and hydrogen powered by AI-driven electrolysis offers a powerful solution. By harnessing the abundant energy of hydrogen and optimizing its production through AI, we can create a cleaner, more sustainable, and equitable global energy system. The integration of hydrogen into renewable energy grids, coupled with advancements in AI technology, has the potential to revolutionize energy production, storage, and distribution, ultimately leading to a carbon-neutral future. With the right investments in research, policy, and infrastructure, AI-driven hydrogen systems can help address the most pressing global challenges of the 21st century, offering a path to a cleaner, more resilient, and more sustainable world.

Chapter 12: The Role of Electrolysis in Clean Energy Systems

The ongoing quest to reduce carbon emissions and mitigate climate change has highlighted the urgent need for clean energy systems that do not rely on fossil fuels. Hydrogen, produced through electrolysis using renewable energy sources, stands at the forefront of this transition. When combined with AI-driven optimization, electrolysis can play a transformative role in clean energy systems, helping to decarbonize industries, support energy storage, and integrate seamlessly with renewable energy sources like solar and wind. This chapter explores how AI-enhanced electrolysis can help reduce carbon footprints, optimize energy storage, and integrate hydrogen into the energy grid, laying the foundation for a sustainable energy future.

How AI-Enhanced Electrolysis Could Help Reduce Carbon Footprints

The most immediate benefit of AI-driven electrolysis is its potential to reduce carbon emissions by enabling the production of hydrogen in a clean, sustainable manner. Hydrogen itself is a carbon-free fuel when used in applications like fuel cells, which only emit water vapor as a byproduct. This makes hydrogen a key player in achieving a carbon-neutral energy future.

1. **Green Hydrogen Production**: Traditional methods of hydrogen production, such as steam methane reforming (SMR), rely on fossil fuels like natural gas, which emit significant amounts of CO_2 during the process. In contrast, electrolysis powered by renewable energy sources, such as wind, solar, or hydropower, produces "green hydrogen"—hydrogen with no carbon emissions. AI can optimize the electrolysis process by adjusting the production parameters in real time to match the availability of renewable energy, ensuring that the system operates at peak efficiency.

   AI-enhanced electrolysis systems can also predict the best times to generate hydrogen based on the availability of renewable energy, such as during periods of high solar or wind output. By aligning hydrogen production with clean energy generation, the process not only avoids carbon emissions but also helps to ensure that renewable energy is used efficiently rather than wasted.

2. **Industrial Decarbonization**: One of the greatest challenges in decarbonizing the global economy is the high carbon intensity of industrial sectors such as steel, cement, and chemicals. These industries are heavily reliant on fossil fuels for energy and feedstock. Hydrogen produced via electrolysis offers a clean alternative to traditional carbon-intensive processes. For instance, hydrogen can replace coke (a carbon-rich material) in the production of steel, helping to significantly reduce emissions in the steel industry.

   By optimizing electrolysis processes with AI, hydrogen production can be made more cost-effective and scalable, allowing industries to adopt hydrogen as a cleaner fuel source. AI's ability to forecast energy needs, optimize energy production, and ensure efficient usage of renewable energy sources will be essential in decarbonizing these hard-to-abate sectors.

3. **Transportation and Mobility**: Hydrogen-powered vehicles, including cars, buses, trucks, and even ships, are emerging as viable alternatives to fossil fuel-powered transportation. AI-driven electrolysis systems can contribute to this shift by ensuring that hydrogen production is aligned with demand. AI can optimize hydrogen refueling stations and distribution networks, enabling seamless integration with transportation infrastructure and helping to accelerate the adoption of hydrogen-powered vehicles.

Moreover, AI can be used to predict the energy consumption patterns of hydrogen-powered transportation, adjusting hydrogen production accordingly to ensure that there is always a sufficient supply of fuel. By creating a more efficient, demand-responsive hydrogen production system, AI-driven electrolysis can support the widespread deployment of hydrogen-based mobility solutions.

Integration with Renewable Energy Sources (Solar, Wind, Hydro)

Renewable energy sources like solar, wind, and hydro are intermittent, meaning they produce energy only when environmental conditions are favorable (i.e., when the sun is shining, the wind is blowing, or water is flowing). This variability can pose challenges to grid stability and energy reliability. Hydrogen, produced through electrolysis, offers a solution by providing a means to store excess renewable energy for later use.

1. **Energy Storage and Load Balancing**: One of the most critical applications of hydrogen in clean energy systems is its role in energy storage. During times of surplus renewable energy generation, AI-driven electrolysis systems can convert excess electricity into hydrogen, which can then be stored in tanks or other storage devices. This hydrogen can be used later to generate electricity when renewable energy production is low or when demand spikes.

   AI can play a key role in managing energy storage systems, ensuring that hydrogen is produced at times when renewable energy output is high and is used efficiently when demand for electricity rises. AI algorithms can predict periods of low energy availability and adjust production schedules for hydrogen storage, optimizing energy supply and demand.

2. **Grid Integration and Smart Grids**: Smart grids, which use advanced sensors, communication systems, and AI algorithms to manage electricity distribution, are essential for integrating renewable energy sources into the grid. AI-driven electrolysis systems can be integrated into these smart grids, enabling hydrogen to be produced in a decentralized manner and delivered to areas of the grid that need it most.

   Hydrogen can also be used as a backup power source to stabilize the grid during periods of high demand or when renewable energy generation is insufficient. By storing excess energy in the form of hydrogen and using AI to manage its distribution, hydrogen can help to balance the grid, ensuring a stable, reliable energy supply.

3. **Microgrids and Decentralized Energy Systems**: Decentralized energy systems, such as microgrids, are becoming increasingly important in rural and off-grid areas. AI-driven electrolysis systems can support these microgrids by providing a local, sustainable energy source. In off-grid communities, for example, renewable energy such as solar or wind can be used to power electrolysis systems that produce hydrogen for storage and later use.

   By integrating hydrogen storage with decentralized energy systems, communities can become more energy-independent and resilient to disruptions in the main grid. AI will be critical in optimizing hydrogen production, ensuring that the system functions efficiently, and monitoring energy needs in real time.

Energy Storage and Distribution Systems Optimized by AI

Energy storage is one of the most significant barriers to fully utilizing renewable energy. AI-driven electrolysis can address this by providing a scalable and efficient means of storing energy in the form of hydrogen, which can be converted back into electricity or used directly as a fuel source.

1. **Optimized Hydrogen Storage**: Hydrogen storage systems must be designed to store hydrogen safely and efficiently, ensuring that it is available when needed. AI can help optimize storage by monitoring pressure, temperature, and other variables, adjusting the conditions under which hydrogen is stored to maximize safety and minimize energy loss.

   AI can also predict when hydrogen will be required for distribution or energy generation, allowing storage systems to be prepared in advance. This helps to avoid supply shortages and ensure that hydrogen is available when demand spikes.

2. **Efficient Distribution Networks**: Distributing hydrogen efficiently requires a network of pipelines, refueling stations, and storage facilities. AI can help optimize these networks, ensuring that hydrogen is transported from production sites to consumption points with minimal energy loss. AI can analyze real-time data on supply and demand, adjusting distribution routes and schedules to ensure timely delivery and reduce inefficiencies.

3. **Integrating Hydrogen with Existing Infrastructure**: For hydrogen to become a mainstream energy source, it must be integrated into existing infrastructure, including power plants, transportation networks, and residential energy systems. AI-driven electrolysis can help facilitate this transition by ensuring that hydrogen is seamlessly incorporated into the broader energy system, enabling a smoother and more cost-effective shift to a hydrogen-based economy.

Conclusion

AI-enhanced electrolysis is poised to play a transformative role in clean energy systems, helping to reduce carbon footprints, optimize energy storage, and integrate hydrogen into the broader energy grid. By enabling the production of green hydrogen using renewable energy sources, AI-driven electrolysis can contribute to global decarbonization efforts, support the transition to renewable energy, and create a more sustainable and resilient energy system.

The integration of hydrogen into clean energy systems is not just a technological challenge, but also an opportunity to rethink how we generate, store, and distribute energy. As we continue to innovate and develop new AI-driven solutions for hydrogen production, we move closer to a future where clean, sustainable energy is accessible to all, and the energy crisis is no longer a barrier to progress.

Chapter 13: The Ethical and Social Responsibility of AI-Driven Electrolysis

As AI-driven electrolysis systems evolve, transforming industries and energy systems, their integration into human society will inevitably raise a host of ethical and social considerations. The ability to harness energy from water using artificial intelligence promises substantial benefits, but it also demands careful thought about the potential consequences for individuals, communities, and the planet. This chapter explores the ethical and social responsibilities associated with AI-driven electrolysis, emphasizing the need for responsible innovation, equitable access, and the careful management of societal impacts.

The Responsibility of Innovators and Leaders to Ensure Positive Societal Impact

Innovation in AI and electrolysis presents an opportunity to address some of the world's most pressing issues, including the energy crisis, climate change, and human health. However, the transformative potential of these technologies carries significant responsibilities. Innovators, researchers, and leaders in the fields of AI and hydrogen energy must ensure that their work aligns with broader societal goals, such as promoting sustainability, reducing inequalities, and fostering social well-being.

1. **Ethical Innovation**: As we develop and deploy AI-driven electrolysis systems, we must prioritize ethical considerations at every stage of the process. This means ensuring that these technologies are designed to benefit society as a whole, not just a select few. Ethical innovation involves considering the long-term impacts of AI on individuals and communities, and striving to create solutions that address social, environmental, and economic inequalities.

For example, while AI can help optimize hydrogen production and distribution, we must ensure that the technologies are not monopolized by a few corporations or countries, but are accessible to those who need them most. The potential for energy systems to become more decentralized and accessible must be fully realized, ensuring that underserved populations in both developed and developing nations can benefit from these innovations.

2. **Accountability and Transparency**: With great power comes great responsibility. As AI algorithms become more integrated into critical systems like energy production, transparency and accountability must be built into their design and operation. Developers and companies must clearly explain how their AI systems work, the data they use, and the decisions they make, particularly when these decisions impact public welfare.

   Moreover, AI systems should be tested and monitored for unintended consequences, such as biases in energy distribution or privacy violations. Ensuring that AI systems remain transparent and accountable to the public is vital for fostering trust in the technology and mitigating concerns about its impact.

3. **Environmental Stewardship**: AI-driven electrolysis systems, when optimized properly, can significantly reduce carbon emissions and contribute to global sustainability efforts. However, this potential will only be realized if innovators prioritize environmental stewardship in every phase of development. From the materials used in electrolysis systems to the energy sources powering them, every aspect of the technology must be assessed for its environmental impact. Companies and researchers must adopt life-cycle assessments to understand and minimize the ecological footprint of their products, ensuring that AI-driven electrolysis remains a sustainable solution in the long term.

Addressing Concerns About the Future of AI Integration into Human Life

The integration of AI-driven technologies into everyday life, especially in the form of bio-energy systems, raises important questions about the future of human autonomy, privacy, and control. As AI systems increasingly interact with biological systems, particularly in medical or human augmentation applications, we must carefully examine the implications for individual rights and freedoms.

1. **Human Autonomy and AI-Powered Systems**: The introduction of AI-powered bio-energy systems—such as implants that harness hydrogen for energy—could enhance human capabilities, but it could also create concerns about human autonomy. If AI systems are embedded in the human body to manage energy production, there may be concerns about control and decision-making. Who controls the AI systems that regulate bio-energy? How much influence does the individual have over these systems?

   To address these concerns, it will be important to establish clear boundaries around the role of AI in human augmentation. Ensuring that individuals maintain control over the systems within their bodies and that AI serves as a supportive, not intrusive, tool is crucial. Furthermore, systems must be designed to allow individuals to opt out, modify, or deactivate AI-driven devices when necessary.

2. **Privacy and Data Security**: The integration of AI into human systems will undoubtedly involve the collection of sensitive data, including information about an individual's health, energy production, and biological processes. The privacy and security of this data must be a top priority, as unauthorized access could lead to significant violations of personal rights.

   Developing strong data protection frameworks and encryption methods will be essential to safeguarding individuals' privacy. Furthermore, individuals must be fully informed about what data is being collected, how it will be used, and who will have access to it. Consent must be a central aspect of AI-powered bio-energy systems, with individuals having the right to control the data generated by these systems.

3. **Equitable Access to Technology**: As with any emerging technology, there is a risk that AI-driven electrolysis systems could exacerbate existing social and economic inequalities. Access to clean, renewable energy, as well as the health benefits associated with AI-driven bio-energy systems, must not be restricted to only the wealthy or technologically advanced nations.

   Governments, international organizations, and private companies must work together to ensure that these technologies are accessible to all, particularly those in underserved regions. Programs that support affordable access to AI-powered energy solutions, as well as partnerships between the public and private sectors, will be key in creating a more equitable global energy landscape.

   Additionally, the adoption of AI-driven electrolysis in health and medical applications should be guided by principles of equity, ensuring that individuals from diverse socioeconomic backgrounds benefit from advancements in personalized medicine and energy systems.

Ensuring Access to Technology and Equitable Benefits for All

The promise of AI-driven electrolysis as a clean, sustainable, and efficient energy solution holds immense potential, but it must be accompanied by a commitment to ensuring that the benefits of these technologies are widely distributed. Achieving this requires collaboration across industries, governments, and communities to develop inclusive policies that prioritize access to technology and reduce barriers to adoption.

1. **Inclusive Policy Development**: Policymakers must develop frameworks that promote inclusive growth and ensure that AI-driven electrolysis benefits all sectors of society. This includes providing financial incentives for renewable energy adoption, funding research into the development of affordable AI-powered electrolysis systems, and fostering international cooperation to support the global transition to clean energy.

   International agreements, such as the Paris Climate Agreement, must be expanded to address the role of hydrogen and AI in achieving global sustainability goals. By incorporating AI-driven electrolysis into these frameworks, nations can work together to mitigate the impacts of climate change and create a fairer, more sustainable world.

2. **Collaborative Efforts Across Sectors**: Collaboration is essential for overcoming the challenges of scaling AI-driven electrolysis systems. Research institutions, private companies, non-governmental organizations (NGOs), and governments must work together to address technological, economic, and societal barriers to adoption. This collaboration will ensure that resources are allocated efficiently, knowledge is shared, and technological advancements benefit the broader public. Partnerships with communities in developing regions can help ensure that AI-driven electrolysis solutions are implemented where they are most needed, helping to address both energy poverty and climate change.

Conclusion

AI-driven electrolysis holds the promise of transforming how we generate, store, and utilize energy, offering clean solutions to some of the most pressing global challenges. However, as we advance these technologies, it is essential to uphold ethical principles that prioritize societal well-being, human rights, and environmental sustainability. The responsibility of innovators, policymakers, and leaders is clear: they must ensure that AI and hydrogen technologies are developed in a way that benefits all of humanity, respects individual autonomy, and contributes to a more just and sustainable world. By embracing these responsibilities, we can harness the full potential of AI-driven electrolysis to power a cleaner, fairer, and more resilient future.

Chapter 14: Conclusion: A New Age of Human Potential

The convergence of artificial intelligence (AI) and electrolysis offers groundbreaking possibilities for human energy systems, with far-reaching implications for sustainability, healthcare, and our very understanding of energy production. As we stand at the cusp of a new age, the ability to harness hydrogenic energy from water using AI-driven electrolysis systems promises to transform the way we live, work, and interact with our environment. This chapter summarizes the transformative potential of this technology and offers a forward-looking perspective on its impact on humanity and the planet.

The Transformative Potential of AI-Driven Electrolysis

AI-driven electrolysis is poised to revolutionize energy production, offering a clean, sustainable alternative to fossil fuels. By leveraging AI's ability to optimize electrolysis processes, we can efficiently produce hydrogen, a clean energy carrier that holds the potential to decarbonize various sectors, from transportation and manufacturing to healthcare and residential energy systems.

Hydrogen produced through electrolysis using renewable energy sources such as wind, solar, and hydropower is poised to become a cornerstone of the global clean energy transition. AI's role in enhancing the efficiency and scalability of this process cannot be overstated. By making electrolysis systems smarter, more adaptive, and more efficient, AI is unlocking hydrogen's true potential as a sustainable energy solution.

Beyond the energy sector, AI-driven electrolysis also holds significant promise for human health. The potential for implantable devices powered by hydrogen could usher in a new era of self-sustaining medical technologies, where energy generation is decentralized and continuous. This technology could lead to advances in personalized medicine, reducing dependency on traditional battery-powered devices, and improving the overall quality of life for individuals with chronic conditions.

A Path to Sustainable Energy for All

The integration of AI with electrolysis is not only about technological innovation but also about making sustainable energy solutions accessible to all. One of the most critical challenges of the modern era is the unequal distribution of energy resources. Nearly 800 million people still lack access to reliable electricity, with much of this disparity occurring in developing regions. AI-powered electrolysis systems have the potential to address this issue by facilitating decentralized hydrogen production, allowing communities to generate and store their own energy locally, without relying on centralized power grids.

Moreover, the scalability of hydrogen production offers a unique opportunity to address the global energy crisis. AI can help optimize energy storage, making it possible to store excess renewable energy in the form of hydrogen for later use, ensuring that energy is available when needed most. With proper implementation, hydrogen-based energy systems could provide reliable, clean, and cost-effective energy solutions, driving socioeconomic development in underserved areas.

However, the challenge lies not only in the development of the technology but in ensuring that it is accessible and equitable. As we advance AI-driven electrolysis, we must prioritize affordability, inclusivity, and accessibility. Efforts must be made to ensure that these technologies are not confined to wealthy nations or industries but are available to those who need them most, particularly in developing countries and marginalized communities.

The Role of AI in Shaping the Future of Energy

AI is uniquely positioned to optimize hydrogen production, making the process more efficient, cost-effective, and adaptable to dynamic environmental conditions. AI-driven systems can predict energy demands, adjust electrolysis processes in real time, and ensure that hydrogen production aligns with the availability of renewable energy. Furthermore, AI can play a key role in integrating hydrogen into the broader energy grid, facilitating the smooth transition to a low-carbon, renewable-based energy system.

AI's role extends beyond optimizing energy production. It can help shape future energy systems by enhancing grid stability, supporting demand-response mechanisms, and enabling better coordination between energy producers and consumers. Smart grids, powered by AI, can automatically adjust to fluctuations in renewable energy generation, seamlessly integrating hydrogen as a key component of the energy network. In doing so, AI-driven electrolysis can help ensure that energy is distributed efficiently, supporting a sustainable, resilient energy infrastructure.

In the coming decades, AI and electrolysis will not only revolutionize the way energy is produced and consumed but also change the very fabric of our energy systems. By enabling decentralized, renewable-based energy production and storage, AI-driven electrolysis could democratize energy access, creating a more sustainable, equitable world for future generations.

Ethical Considerations and Societal Responsibility

As with any new technology, the development and deployment of AI-driven electrolysis must be guided by strong ethical frameworks. While the potential benefits are vast, we must carefully consider the social, economic, and environmental implications of this technology. Ethical considerations must include ensuring privacy, safety, and informed consent in AI-powered medical applications, as well as addressing potential risks associated with human augmentation.

Moreover, we must be mindful of the impact on jobs and industries that may be disrupted by the widespread adoption of AI-driven electrolysis systems. Efforts must be made to ensure that workers are retrained and integrated into the emerging green economy, where new jobs and opportunities in renewable energy, hydrogen production, and AI development will be created.

The rapid pace of technological advancement demands that we take a responsible approach to innovation. Policymakers, business leaders, and researchers must collaborate to ensure that AI-driven electrolysis technologies are developed in a way that benefits society as a whole, promotes sustainability, and addresses the needs of vulnerable populations. Access to technology must be equitable, and its benefits must be widely distributed to avoid exacerbating existing inequalities.

A Vision for the Future: Sustainability, Health, and Empowerment

The integration of AI-driven electrolysis into energy systems, medical technologies, and everyday products signals the dawn of a new era. It is an era where clean energy is abundant, health outcomes are improved through self-sustaining medical devices, and communities have the tools they need to thrive sustainably.

Looking ahead, the possibilities are limitless. As AI-driven electrolysis systems evolve, they will become increasingly integrated into every aspect of human life—from the energy we consume to the medical devices that sustain us. The potential to create a cleaner, more sustainable future has never been more within our reach, but it will require a collective effort to ensure that these innovations are used responsibly, ethically, and equitably.

By focusing on responsible development, equitable access, and environmental stewardship, we can unlock the full potential of AI-driven electrolysis to create a world that is more sustainable, healthy, and empowered. The path to a cleaner, more prosperous future lies in our hands, and with the right vision, commitment, and collaboration, we can achieve a new age of human potential.

Final Thoughts on the Evolution of Energy Systems, Humanity, and AI

As we look toward the future, AI-driven electrolysis represents a pivotal step in the evolution of energy systems and humanity's relationship with energy. The fusion of artificial intelligence and electrolysis holds the key to addressing some of the most pressing challenges of our time—energy sustainability, climate change, and health disparities.

This book has explored the myriad ways in which AI and electrolysis can be harnessed to shape a better future. It is now up to innovators, leaders, and policymakers to continue pushing the boundaries of what is possible, ensuring that these technologies are used to promote positive societal change and environmental preservation. The journey toward a sustainable, AI-driven energy future has just begun, and with collective action, we can unlock a new age of human potential—one that is powered by clean, renewable, and accessible energy for all.

Chapter 15: The Path Forward: What's Next in AI and Hydrogen Energy?

As we look to the future, the intersection of AI and hydrogen energy is poised to play a transformative role in shaping global energy systems. The integration of advanced technologies like AI-driven electrolysis will not only enable cleaner energy production but also revolutionize how we use energy across sectors, from transportation to healthcare. However, as with any technological leap, there are significant challenges to overcome before AI-powered hydrogen systems can be adopted at scale. This chapter explores what lies ahead in the development of AI-driven hydrogen energy, the emerging trends and technologies that could accelerate progress, and the exciting opportunities that await.

Projections for the Future of AI, Electrolysis, and Hydrogen Energy

The next few decades will likely witness rapid advancements in AI, electrolysis, and hydrogen energy. Each of these fields is evolving independently, but together, they form a powerful synergy that can drive the global transition to clean energy. Here are several key trends and projections for the future:

1. **Efficiency and Cost Reduction**: One of the primary goals for the future of AI-driven electrolysis is improving efficiency and reducing costs. As AI algorithms optimize the electrolysis process, hydrogen production costs will decrease, making it a more viable alternative to fossil fuels. AI can adjust parameters in real time to ensure that electrolysis is always operating under optimal conditions, reducing energy consumption and increasing hydrogen output.

   Additionally, advancements in AI-powered material science will lead to the development of more efficient electrolysis catalysts, which will further reduce energy costs. In the long run, this will make hydrogen a competitive option for industries, transportation, and power generation.

2. **Expansion of Hydrogen Infrastructure**: For hydrogen energy to become a mainstream solution, a global hydrogen infrastructure must be established. This includes hydrogen production facilities, storage systems, distribution networks, and refueling stations for hydrogen-powered vehicles. AI can play a crucial role in managing these networks, optimizing energy distribution, and ensuring that hydrogen is delivered where and when it is needed most.

   Over the next few decades, we expect to see the growth of hydrogen hubs, particularly in regions with abundant renewable energy resources, such as wind and solar farms. AI-driven systems will help manage the decentralized production and distribution of hydrogen, enabling the creation of a resilient, interconnected energy infrastructure that serves both urban and rural areas.

3. **Integration with Smart Grids and Energy Systems**: The integration of AI-driven hydrogen energy systems into smart grids will be another key development in the coming years. Smart grids, which use AI to balance supply and demand, will be able to incorporate hydrogen as a flexible energy storage and dispatch system. Hydrogen can serve as a buffer for renewable energy sources, smoothing out fluctuations in energy generation from solar and wind.

By 2050, we may see widespread adoption of hydrogen as a key component of energy systems, complementing electricity from renewables and acting as a backup for grid stability. AI will manage the coordination between these energy sources, ensuring that the right mix of energy is available at all times, reducing reliance on traditional fossil fuels.

4. **Hydrogen in Transportation**: Hydrogen-powered vehicles, including trucks, trains, buses, and even ships, are expected to become a significant part of the transportation landscape. The growth of hydrogen infrastructure will be essential to making this transition possible, and AI will be central to managing refueling stations, optimizing vehicle energy consumption, and improving the overall efficiency of hydrogen transport systems.

Within the next 10 to 20 years, we can expect to see more widespread deployment of hydrogen fuel cell vehicles in urban areas, particularly for heavy-duty applications where battery-electric vehicles are less practical. By leveraging AI, these vehicles will become increasingly efficient, with AI models predicting energy needs and adjusting fuel consumption based on driving patterns, weather conditions, and other factors.

Emerging Trends and Technologies that Could Accelerate Progress

While AI-driven electrolysis and hydrogen energy systems show great promise, there are several emerging trends and technologies that could accelerate their development and adoption:

1. **Advancements in AI and Machine Learning**: As AI continues to evolve, we can expect more sophisticated machine learning algorithms that will enable even more efficient optimization of electrolysis systems. AI models will become better at predicting energy consumption patterns, identifying inefficiencies, and adapting to changes in environmental conditions. These advances will be crucial in scaling hydrogen production and reducing costs.

2. **Quantum Computing**: Quantum computing, still in its early stages, holds the potential to revolutionize hydrogen production by dramatically increasing computational power. With quantum algorithms, researchers could simulate the behavior of molecules and materials at an atomic level, allowing for the design of more efficient catalysts for electrolysis and improving the overall efficiency of hydrogen production. While quantum computing is still far from widespread adoption, it could play a pivotal role in the future of AI-driven electrolysis.

3. **Breakthroughs in Hydrogen Storage**: Efficient hydrogen storage remains one of the key challenges for the widespread adoption of hydrogen as a fuel. Current storage methods, such as high-pressure tanks or liquid hydrogen, are expensive and energy-intensive. Researchers are exploring new materials for hydrogen storage, including solid-state hydrogen storage systems, which could be more compact, safe, and cost-effective.

AI will be instrumental in accelerating these breakthroughs by simulating and optimizing new materials for hydrogen storage. This could lead to the development of more efficient storage solutions that allow hydrogen to be transported more easily and safely, making it a viable fuel for a variety of applications.

4. **Integration with Other Green Technologies**: The future of AI-driven electrolysis and hydrogen energy is not just about standalone technologies but about integrating hydrogen production into the broader ecosystem of green technologies. For example, hydrogen could be used as a storage medium for excess energy produced by wind and solar farms, helping to stabilize the grid and ensure reliable energy access.

   Additionally, AI can help manage the integration of hydrogen with other sustainable technologies, such as carbon capture and storage (CCS) or bioenergy systems. These integrated systems could offer a comprehensive solution for achieving net-zero emissions and creating a circular, sustainable energy economy.

Opportunities and Challenges for the Future

As AI-driven hydrogen energy systems continue to develop, they will present numerous opportunities for innovation and growth. However, several challenges must be addressed to ensure that this technology can scale globally.

1. **Scaling for Global Adoption**: One of the major challenges will be scaling AI-driven electrolysis systems to meet the energy demands of entire nations or industries. While small-scale hydrogen production systems are already being deployed, scaling these technologies to meet the global demand for clean energy will require substantial investments in infrastructure, research, and development.

2. **Public and Private Sector Collaboration**: The path forward will require close collaboration between governments, private companies, and international organizations. Governments must provide supportive policies and incentives for the development of hydrogen infrastructure and the adoption of clean technologies. Private companies, on the other hand, will be responsible for driving innovation and reducing costs to make hydrogen energy affordable and accessible.

3. **Regulatory and Safety Concerns**: As hydrogen energy becomes more widespread, regulatory frameworks must be developed to ensure the safe production, storage, and transportation of hydrogen. AI-driven electrolysis systems must meet stringent safety standards to prevent accidents and ensure the integrity of the entire hydrogen supply chain.

4. **Public Perception and Education**: Widespread adoption of hydrogen energy will also require a shift in public perception. Hydrogen is still viewed with some skepticism due to concerns about safety and its cost. Public education campaigns will be essential in informing consumers about the benefits of hydrogen and its role in creating a sustainable energy future.

Conclusion

The path forward for AI-driven electrolysis and hydrogen energy is both exciting and challenging. While there are significant hurdles to overcome, the potential benefits for the global energy landscape are vast. As we continue to refine the technologies, collaborate across sectors, and overcome barriers to adoption, hydrogen energy powered by AI-driven electrolysis will play a central role in building a clean, sustainable, and resilient future.

The coming decades offer a unique opportunity to reshape the world's energy systems, and with the continued convergence of AI, renewable energy, and hydrogen, the possibilities are limitless. Through innovation, collaboration, and responsible development, we can unlock the true potential of hydrogen energy, ushering in a new era of clean, efficient, and equitable energy for all.

Chapter 16: Conclusion: A New Age of Human Potential

The fusion of artificial intelligence (AI) with electrolysis technology represents a powerful leap forward in the quest to unlock hydrogenic energy from water, presenting humanity with groundbreaking possibilities for both energy production and medical advancements. As we embark on this transformative journey, AI-driven electrolysis not only holds the potential to revolutionize the global energy system, but it also invites a new era of human capabilities, sustainability, and health.

In this final chapter, we reflect on the immense potential of AI-driven electrolysis, summarize the possibilities and challenges that lie ahead, and explore the role of this emerging technology in shaping a cleaner, more efficient, and sustainable future.

The Power of AI-Driven Electrolysis: Unlocking New Possibilities

AI and electrolysis together have created an ecosystem that can significantly reduce carbon footprints, provide abundant clean energy, and even enhance human health. As hydrogen energy systems powered by AI evolve, they promise to reshape multiple industries, including transportation, healthcare, manufacturing, and beyond. Through efficient and sustainable hydrogen production, these systems address the urgent need for clean energy, offering a solution to the world's dependence on fossil fuels.

1. **Revolutionizing Energy Systems**: The ability to generate hydrogen from water using renewable energy, guided by AI-optimized electrolysis, provides a significant leap forward in solving the global energy crisis. AI enhances electrolysis by continuously adjusting operational parameters to maximize efficiency, ensuring that hydrogen can be produced, stored, and distributed on a large scale. This shift to a hydrogen-based energy economy, integrated into the smart grids of the future, could lead to a cleaner, more resilient, and decentralized global energy infrastructure.
2. **Healthcare Advancements**: In addition to its role in energy systems, AI-driven electrolysis opens new frontiers in healthcare. The potential for implantable devices that harness hydrogen for energy could eliminate the need for traditional battery systems in medical implants, providing self-sustaining, long-lasting power sources for devices like pacemakers and neurostimulators. As AI assists in managing the bio-energy systems in the human body, it could lead to breakthroughs in personalized medicine, empowering humans to optimize their health while reducing their dependency on external energy sources.

3. **Enhancing Sustainability**: Beyond hydrogen production, AI-driven electrolysis is key to sustainability. By making hydrogen a cost-competitive, green alternative to traditional energy sources, we can achieve significant reductions in carbon emissions. As industries such as transportation, power generation, and manufacturing transition to hydrogen, we take one step closer to realizing the vision of a zero-carbon economy. The environmental impact of hydrogen-based solutions, combined with AI's ability to optimize processes, will contribute to a more circular economy, where resources are used efficiently and waste is minimized.

Challenges and Barriers to Overcome

While the promise of AI-driven electrolysis is immense, the journey toward widespread adoption and realization of its full potential comes with challenges. Scaling this technology, ensuring its ethical application, and overcoming the economic and technological barriers that remain will require significant effort and collaboration from all sectors of society.

1. **Technological Barriers**: The efficiency of electrolysis processes must be further improved to make hydrogen a viable large-scale energy source. AI-driven electrolysis systems must be refined to maximize output while minimizing energy input. Additionally, breakthroughs in hydrogen storage, transportation, and integration with existing infrastructure are essential for hydrogen to become a mainstream energy source. AI will play a key role in optimizing these systems, but continued research and development in materials science, nanotechnology, and energy storage will be necessary.
2. **Economic and Practical Barriers**: The initial cost of developing and scaling AI-powered electrolysis systems, along with the infrastructure needed to support hydrogen production and storage, can be significant. As with any new technology, financial investments and policy support are essential to reducing costs and promoting widespread adoption. Governments, industry leaders, and international organizations must work together to fund research and create policies that promote the growth of hydrogen infrastructure while ensuring affordability for consumers.

3. **Social and Ethical Considerations**: The integration of AI into human biology, particularly in medical applications, raises ethical concerns about privacy, consent, and the potential for human augmentation. As AI-driven electrolysis systems are incorporated into human-centered applications, ensuring transparency, informed consent, and data security will be essential. Moreover, as the technology advances, we must carefully consider the broader societal implications of AI-powered systems—how these technologies could affect employment, individual autonomy, and access to healthcare.

4. **Geopolitical and Regulatory Challenges**: The global adoption of AI-driven hydrogen technologies will also require significant geopolitical cooperation. Hydrogen, as a global resource, will affect international trade, supply chains, and energy security. Effective regulations, standards, and international agreements must be established to govern the production, distribution, and use of hydrogen. These global frameworks will ensure that hydrogen energy is used to benefit all nations and communities, especially those in developing regions, while also preventing monopolies from limiting access to the technology.

The Role of Collaboration in Shaping the Future

The success of AI-driven electrolysis technology depends on collaboration across industries, governments, researchers, and communities. No single entity can tackle the challenges of scaling this technology, developing infrastructure, or ensuring equitable access to its benefits. Collaborative efforts are essential for unlocking the full potential of AI-powered hydrogen systems.

1. **Interdisciplinary Collaboration**: Breakthroughs in AI, electrolysis, and hydrogen technologies will come from interdisciplinary collaboration. Partnerships between AI experts, bioengineers, energy researchers, policymakers, and the private sector will drive the innovation needed to solve technical challenges and scale the technology for global use. Cross-sector collaborations will also ensure that AI-driven electrolysis is deployed in ways that are socially responsible and equitable.

2. **Global Cooperation**: The transition to a global hydrogen economy requires international cooperation. Hydrogen production and storage infrastructure must be developed worldwide, and knowledge sharing between nations will be critical. By pooling resources and knowledge, countries can accelerate the development of hydrogen technologies and create global markets for clean energy solutions. International policies and agreements will also be necessary to promote the responsible use of AI in energy systems and prevent its misuse.

3. **Public and Private Partnerships**: Governments, industries, and research institutions must form strong partnerships to promote innovation, fund research, and develop the necessary infrastructure for large-scale adoption. Public-private partnerships will be crucial for overcoming the initial cost barriers and ensuring that AI-driven hydrogen solutions are affordable and accessible to all. In addition, ensuring that the benefits of this technology are distributed equitably will require active engagement with communities and stakeholders around the world.

A Vision for a Clean, Sustainable Future

Looking to the future, AI-driven electrolysis holds the promise of a cleaner, more sustainable world, where energy is abundant, accessible, and efficient. As this technology continues to evolve, it will redefine the way we produce, store, and consume energy, driving us closer to a net-zero carbon economy. It will also empower individuals, communities, and industries to take charge of their energy needs, promoting energy independence and resilience in the face of global challenges.

In healthcare, AI-driven electrolysis will enable life-saving innovations, improving the quality of life for millions through self-sustaining, energy-efficient medical devices. At the same time, the widespread adoption of hydrogen energy will mitigate the environmental impact of industries, leading to cleaner air, healthier communities, and a more sustainable planet.

The evolution of AI-driven electrolysis is not just about technology; it's about harnessing human potential to create a better, more sustainable world. By embracing this technology and its ethical implications, we are not only solving the energy crisis but also reshaping the future of humanity. Together, we have the opportunity to shape a world where energy, health, and the environment are in harmony—an era of clean, sustainable, and limitless possibilities.

Final Thoughts

The journey to unlock the full potential of AI-driven electrolysis and hydrogen energy has only just begun. While challenges remain, the progress made thus far offers hope and inspiration for what lies ahead. By continuing to innovate, collaborate, and think boldly, we can move toward a future where energy is clean, abundant, and accessible to all, and where AI plays a central role in achieving this vision. The future of human energy is here, and with it, a new age of human potential—powered by the convergence of technology, nature, and ingenuity.

Chapter 17: The Role of Research and Innovation in Shaping the Future

The journey to harnessing the full potential of AI-driven electrolysis is not only a technological endeavor, but a call for continuous innovation and interdisciplinary research. The future of hydrogen energy and human integration with AI-powered electrolysis depends on the collective contributions of scientists, engineers, policymakers, and innovators. The role of research in advancing electrolysis, AI, and the fusion of these technologies is critical to overcoming existing barriers, improving efficiency, and ensuring that these technologies are scalable and accessible to all.

In this chapter, we explore the importance of research and innovation in advancing AI-driven electrolysis, the role of interdisciplinary collaboration in fostering breakthroughs, and how future education and innovation will shape the bio-energy technologies of tomorrow.

The Importance of Scientific Research in Advancing Electrolysis and AI Technologies

At the heart of every technological advancement is scientific research. The development of AI-driven electrolysis systems is no different. To make these systems viable for large-scale use, continued research is needed in several key areas:

1. **Electrolysis Efficiency**: One of the primary goals of research in electrolysis is improving efficiency. Currently, electrolysis requires significant energy input, and making this process more energy-efficient is essential for reducing costs and making hydrogen a competitive alternative to fossil fuels. Research into new catalysts, advanced materials, and optimized electrolysis techniques will be crucial in achieving this goal.

2. **AI Optimization Algorithms**: AI's role in electrolysis extends far beyond automating the process. AI can optimize electrolysis systems by analyzing vast datasets, predicting energy requirements, and adjusting parameters in real time to maximize efficiency. The development of more advanced machine learning models will enable AI to become a more powerful tool in fine-tuning the process of hydrogen production, ensuring that it operates efficiently under varying conditions.

3. **Hydrogen Storage and Transportation**: Once hydrogen is produced, it needs to be stored and transported efficiently. Current methods of hydrogen storage, such as pressurized tanks or liquid hydrogen, present significant challenges in terms of safety, cost, and energy consumption. Continued research into alternative storage methods, such as solid-state hydrogen storage, will play a critical role in the widespread adoption of hydrogen as a clean energy source. AI will assist in developing and optimizing these storage methods to ensure their scalability and cost-effectiveness.

4. **Integration with Renewable Energy Sources**: The success of AI-driven electrolysis systems depends on their ability to integrate seamlessly with renewable energy sources such as wind and solar power. AI can optimize the operation of electrolysis systems by predicting when renewable energy sources will be abundant and when they will be scarce, adjusting hydrogen production accordingly. Research into energy grid optimization, smart grids, and energy storage will be key to ensuring that hydrogen can be produced reliably and efficiently, even when renewable energy generation fluctuates.

Interdisciplinary Collaboration: The Key to Breakthroughs

Innovation in AI-driven electrolysis requires collaboration across multiple disciplines. The intersection of AI, renewable energy, electrolysis, materials science, bioengineering, and environmental science means that advancements will not come from any single field in isolation. Interdisciplinary collaboration is critical for addressing the complex challenges that arise in scaling this technology and finding solutions to its technical and economic barriers.

1. **Collaboration Between AI and Engineering**: AI and engineering must work in tandem to develop practical, real-world electrolysis systems. Engineers provide the practical knowledge of how systems work and how they can be optimized, while AI experts create algorithms that can adapt these systems to new conditions. This collaboration ensures that electrolysis systems not only work in theory but are also scalable and efficient in real-world applications.

2. **Bioengineering and AI Integration**: The potential for human integration with AI-powered electrolysis systems lies at the intersection of bioengineering and artificial intelligence. Scientists working in bioengineering will need to collaborate with AI experts to design systems that are compatible with human biology, ensuring that AI-powered devices can work seamlessly within the human body. Innovations in biotechnology, tissue compatibility, and bio-energy systems will be needed to bring these concepts to life.

3. **Environmental and Sustainability Research**: As AI-driven electrolysis technologies are deployed, their environmental impact must be considered. Research into sustainable practices, lifecycle assessments, and the reduction of environmental footprints will ensure that hydrogen production remains environmentally beneficial. Sustainable sourcing of materials, waste management, and resource efficiency must be addressed to ensure that AI-driven electrolysis contributes positively to the circular economy and global sustainability.

4. **Public Policy and Regulatory Research**: Alongside technical advancements, the successful integration of AI-driven electrolysis into global energy systems requires careful research into policy, regulation, and governance. Developing regulatory frameworks that ensure safety, equity, and environmental protection is essential as hydrogen production becomes more widespread. Policymakers and researchers must collaborate to create standards that guide the adoption of hydrogen technologies and ensure that they are used responsibly and equitably.

The Role of Education and Innovation in Bio-Energy Technologies

As AI and electrolysis technologies continue to evolve, the next generation of scientists, engineers, and policymakers must be equipped with the knowledge and skills necessary to drive these advancements forward. The role of education in fostering innovation and ensuring the sustainable development of bio-energy technologies cannot be overstated.

1. **Fostering Cross-Disciplinary Education**: To meet the challenges of developing AI-driven electrolysis systems, educational programs must foster cross-disciplinary collaboration. Universities and research institutions should offer programs that combine AI, engineering, renewable energy, bioengineering, and sustainability. Encouraging students from diverse fields to work together will help ensure that solutions are developed holistically and that innovations are informed by a variety of perspectives.

2. **Supporting Innovation and Entrepreneurial Thinking**: Innovation thrives in environments where creativity and entrepreneurship are encouraged. Researchers and institutions should promote entrepreneurial thinking, enabling individuals to explore novel applications for AI-driven electrolysis technologies and hydrogen energy. Supporting startup ecosystems, incubators, and innovation hubs will drive new ideas and accelerate the commercialization of hydrogen technologies.

3. **Public Awareness and Training**: As hydrogen energy and AI-powered systems become more prevalent, it is essential to raise public awareness and educate communities about the benefits of these technologies. Training programs for workers in the energy, healthcare, and technology sectors will ensure that the workforce is prepared for the shifts brought about by AI and hydrogen energy. Public education campaigns can help demystify the technology and encourage its acceptance and adoption.

Shaping the Future: A Global Effort

Advancing AI-driven electrolysis and hydrogen energy is a global effort that will require the collaboration of governments, industries, academic institutions, and communities worldwide. The challenges are significant, but the potential benefits are vast. With continued research, interdisciplinary collaboration, and a commitment to innovation, AI-driven electrolysis can play a critical role in achieving a sustainable, low-carbon future for the planet.

As we move forward, the road to a sustainable, hydrogen-powered world requires vision, determination, and a unified global effort. By fostering collaboration across disciplines and ensuring that research and innovation are prioritized, we can pave the way for the next generation of energy solutions and realize the full potential of AI-driven electrolysis. Through collective action, we can turn these possibilities into reality, shaping a future where energy is cleaner, more sustainable, and accessible for all.

Chapter 18: Pioneers and Visionaries: Leaders in AI and Hydrogen Technologies

As AI-driven electrolysis and hydrogen technologies continue to evolve, visionary leaders and pioneers are shaping the future of energy production, medical technologies, and the broader integration of artificial intelligence in sustainable systems. These individuals—ranging from scientists and engineers to entrepreneurs and innovators—are pushing the boundaries of what is possible, fostering interdisciplinary collaboration, and driving advancements that will fundamentally alter how we think about energy, sustainability, and human potential.

This chapter explores some of the key figures and organizations leading the way in AI and hydrogen energy. It highlights their contributions to the emerging fields of AI-driven electrolysis and bioenergy, showcasing how their work is contributing to the realization of a cleaner, more sustainable world.

Profiles of Leading Scientists, Engineers, and Innovators in the Field

1. **Dr. Jennifer Doudna** – Co-creator of CRISPR Gene Editing: While her work is primarily focused on genetics, Dr. Jennifer Doudna's pioneering contributions to biotechnology have opened new avenues for bioengineering, one of the key areas influencing AI-driven electrolysis systems. With her revolutionary CRISPR technology, Dr. Doudna has provided scientists with the tools to edit genes, potentially enabling the creation of bioengineered systems capable of optimizing energy production at the cellular level. Her work in gene editing could lay the groundwork for future advances in bio-energy integration, where biological systems and artificial intelligence converge.

2. **Elon Musk** – Founder of Tesla and SpaceX: Elon Musk's contributions to renewable energy through Tesla's solar and battery technologies are well known. His focus on sustainable energy has extended to developing hydrogen-powered technologies, especially through ventures like SpaceX's development of advanced rocket systems that use hydrogen. Musk's vision of a sustainable, carbon-neutral future through energy innovation has made him a key figure in the development of hydrogen and AI-based energy systems. Through Tesla's energy division, Musk has set the groundwork for AI-driven energy storage solutions and renewable energy distribution.

3. **Dr. Daniel Nocera** – Professor of Energy at Harvard University: Dr. Nocera is a leading researcher in the field of artificial photosynthesis and hydrogen fuel cells. His work focuses on developing a more efficient and cost-effective method of using water to produce hydrogen fuel. Through his groundbreaking work in artificial photosynthesis, Nocera aims to mimic the natural process by which plants convert sunlight into energy. The development of these technologies, powered by AI to optimize energy production, could one day lead to sustainable, renewable hydrogen production at an industrial scale.

4. **Dr. Frances Arnold** – Nobel Laureate in Chemistry: A pioneer in the field of bioengineering, Dr. Frances Arnold's work focuses on enzyme engineering—an area crucial for optimizing electrolysis systems. Enzymes play a key role in catalyzing chemical reactions, including those in electrolysis processes that break water molecules into hydrogen and oxygen. Through directed evolution of enzymes, Arnold has contributed to making these processes more efficient and cost-effective. Her innovative approach could be integrated with AI-driven electrolysis to improve reaction rates, efficiency, and scalability.

5. **John B. Goodenough** – Pioneer in Energy Storage Technology: As the inventor of the lithium-ion battery, John B. Goodenough has made contributions that have had a lasting impact on energy storage solutions, which will be critical for the successful implementation of hydrogen energy systems. Goodenough's research continues to push the boundaries of energy storage, with a focus on creating better, more efficient systems for storing hydrogen and other forms of renewable energy. His work in improving storage capacity and efficiency directly complements the AI-driven electrolysis systems that will rely on effective energy storage solutions for widespread adoption.

6. **The International Hydrogen Energy Agency (IHEA)**: The IHEA is a collaborative body that brings together researchers, innovators, and policymakers to accelerate the adoption of hydrogen as a clean energy source. With a focus on research and development, the agency plays a crucial role in advancing AI-driven electrolysis technologies, promoting international standards, and fostering partnerships that help bridge the gap between hydrogen production and global energy solutions. By coordinating the global efforts to develop hydrogen infrastructure and policy, the IHEA is helping to ensure the widespread adoption of hydrogen as a key pillar of the future energy mix.

How These Pioneers Are Shaping the Future of AI and Energy Systems

1. **AI-Optimized Hydrogen Production**: Innovators in AI and hydrogen energy are exploring ways to use machine learning to enhance the efficiency of electrolysis systems. By using AI models to predict energy consumption and optimize system parameters, researchers and engineers are making hydrogen production more reliable, cost-effective, and scalable. AI's role in predicting renewable energy generation patterns, such as solar and wind power, allows for better integration of hydrogen production with these sources. This synergy helps avoid the overproduction of hydrogen during times of low demand, improving storage and minimizing waste.
2. **AI and Bioengineering Synergy**: The convergence of AI and bioengineering is key to developing advanced bio-energy systems. AI can be used to optimize biological processes at the cellular and molecular level, such as improving the efficiency of enzymes used in electrolysis. Researchers are also exploring ways to bioengineer microbes or plants to aid in hydrogen production, creating a more sustainable and environmentally friendly alternative to current methods. The integration of AI with bioengineering will be critical in realizing a future where hydrogen production is as efficient and cost-effective as possible.

3. **Hydrogen as a Clean and Sustainable Energy Source**: Pioneers in the field are making significant strides in proving that hydrogen can play a major role in decarbonizing industries. From heavy-duty transportation to power generation, hydrogen is emerging as a versatile energy carrier that can replace fossil fuels. AI is playing a crucial role in optimizing hydrogen energy systems, including storage, transportation, and utilization. Through the integration of renewable energy sources and AI-driven electrolysis systems, hydrogen can be produced sustainably, stored, and distributed in an efficient, low-cost manner.

4. **Impact on Global Policy and Infrastructure**: Innovators and organizations like the International Hydrogen Energy Agency are working to establish policies and infrastructure that support the widespread adoption of hydrogen. These efforts will require both governmental and corporate investments in hydrogen infrastructure, including refueling stations, pipelines, and large-scale storage systems. AI will help optimize these systems, ensuring efficient energy distribution and facilitating the smooth integration of hydrogen into the global energy grid.

5. **Driving the Circular Economy**: The shift to AI-driven hydrogen production is also driving the transition to a more circular economy, where waste products are minimized, and resources are used more efficiently. Hydrogen energy systems can be integrated with carbon capture and utilization technologies to create closed-loop systems that reduce emissions. Innovators are also exploring ways to make hydrogen production more sustainable by using waste products as inputs. AI-driven electrolysis systems will be a key enabler of these technologies, allowing them to function at scale and contribute to a more sustainable global economy.

Contributions to Bioenergy and the Emerging AI-Driven Electrolysis Fields

Leaders in AI and hydrogen technologies are not only advancing their fields but also pushing the boundaries of what is possible in bioenergy. By harnessing the power of AI to enhance biological processes and integrating them with advanced electrolysis systems, they are paving the way for new, sustainable energy sources. These contributions are helping to address global challenges such as energy scarcity, climate change, and environmental degradation.

Through their work, these pioneers are creating the foundation for a new energy landscape—one in which AI plays an integral role in optimizing energy production, storage, and consumption. The shift towards AI-driven electrolysis and hydrogen energy will provide humanity with the tools to build a sustainable, zero-carbon future that benefits everyone.

Conclusion

The pioneers and visionaries driving the development of AI and hydrogen technologies are shaping the future of energy, health, and sustainability. Their work is not just about creating innovative technologies, but about pushing the boundaries of what is possible and providing humanity with the tools to address the world's most pressing challenges.

As AI-driven electrolysis continues to evolve, the contributions of these innovators will be felt across industries and communities, transforming the way we think about energy production, consumption, and integration. Their leadership and vision will ensure that the promise of clean, sustainable hydrogen energy becomes a reality—ushering in a new era of technological innovation, environmental stewardship, and human potential.

Chapter 19: The Ethical and Social Responsibility of AI-Driven Electrolysis

As AI-driven electrolysis technologies emerge as a cornerstone for clean, sustainable energy, they also bring with them a host of ethical and social responsibilities. These technologies, while offering remarkable potential to transform energy production and human health, must be deployed with a clear understanding of their societal impact. The decisions we make today regarding the development, regulation, and use of AI in electrolysis will shape not only the future of energy but the future of our society as a whole.

This chapter delves into the ethical and social dimensions of AI-driven electrolysis, exploring the responsibility of innovators, researchers, policymakers, and corporations to ensure that these technologies benefit society in equitable, ethical, and sustainable ways.

The Responsibility of Innovators and Leaders

The development and application of AI-driven electrolysis systems are largely driven by innovators and leaders in technology, energy, and healthcare. These individuals and organizations bear significant responsibility in ensuring that the benefits of these technologies are maximized while minimizing potential risks.

1. **Ensuring Equitable Access**: One of the primary ethical challenges associated with AI-driven electrolysis is ensuring equitable access to the technology. As hydrogen energy systems become more prevalent, it is essential to ensure that they are available not only to affluent nations and communities but also to those in developing regions. Global energy inequalities must be addressed to prevent further marginalization of underdeveloped regions.

   Innovators must focus on making AI-powered hydrogen technologies affordable and accessible to all. Governments and international organizations should work together to create policies and programs that ensure fair distribution, especially in regions that currently lack access to clean energy sources.

2. **Inclusive Development**: Leaders in the field must recognize the importance of inclusivity in technological development. As AI plays an increasingly central role in energy systems, it is essential to ensure that diverse voices—across gender, race, economic background, and geography—are included in the design, implementation, and regulation of these systems. A narrow focus on the needs and experiences of only certain groups may lead to technological disparities that further entrench social inequalities.

3. **Transparency and Accountability**: The use of AI in energy systems raises concerns about transparency and accountability. AI models used to optimize electrolysis and hydrogen production must be designed to be transparent and explainable. This is especially important when AI algorithms influence decisions that affect people's livelihoods, health, and the environment. Organizations must prioritize openness in how AI systems are developed and deployed, ensuring that their decision-making processes are understandable to the public and regulatory bodies.

   Furthermore, accountability mechanisms should be in place to ensure that when things go wrong—whether due to AI system malfunctions, mismanagement, or unethical practices—there are clear paths to accountability. Innovators and leaders must uphold high standards of integrity and responsibility in every phase of the development process.

Addressing Concerns About the Future of AI Integration into Human Life

The integration of AI into human life, particularly in areas like healthcare and bioenergy, raises complex ethical questions about autonomy, privacy, and the nature of human identity. AI-driven electrolysis systems, especially in medical applications such as implantable devices, could provide extraordinary benefits but must also be carefully regulated to protect human rights and dignity.

1. **Autonomy and Human Agency**: The use of AI in healthcare, particularly in implantable devices powered by hydrogen, raises questions about individual autonomy and the right to control one's body. As bioengineering technologies advance, there may be concerns about the extent to which individuals can opt out of these systems or whether they might be coerced or influenced into using them. Ethical guidelines should be established to protect human autonomy, ensuring that individuals have the right to make informed choices about the integration of AI with their biology. Informed consent will be essential, particularly in cases where AI technologies are used to augment human abilities or alter biological processes.

2. **Privacy and Data Protection**: The integration of AI and bioenergy systems, especially in healthcare, will require the collection and analysis of vast amounts of personal data. This data could include health information, biological metrics, and other sensitive personal details that are necessary to ensure the safe and effective functioning of AI-powered medical devices. However, the collection and storage of such data raise significant privacy concerns.

Protecting the privacy of individuals whose data is used by AI systems is a critical responsibility. Regulations such as the General Data Protection Regulation (GDPR) in Europe should be applied to ensure that personal data is stored securely and that individuals' privacy rights are respected. Policies must be put in place to ensure that the data used to optimize AI systems is protected from misuse, unauthorized access, or exploitation.

3. **Safety and Long-Term Impact**: As AI and electrolysis systems are developed for human integration, ensuring the safety and well-being of individuals using these technologies will be paramount. AI-driven medical devices, for example, must be rigorously tested for safety and efficacy, with safeguards in place to prevent malfunctions or harmful side effects. Additionally, long-term impact studies will be necessary to assess the health and societal effects of using AI-driven systems in humans over extended periods.

   The social implications of human augmentation and AI integration must also be considered. As these technologies advance, questions about what it means to be human will inevitably arise. Policymakers must work alongside ethicists, sociologists, and technologists to ensure that the development of these systems does not erode human dignity or create new forms of inequality.

Ensuring Access to Technology and Equitable Benefits for All

For AI-driven electrolysis technologies to be a truly transformative force for good, they must benefit all sectors of society, especially the most vulnerable. Achieving this requires careful consideration of the social and economic factors that can influence the equitable distribution of technology.

1. **Fostering Global Collaboration**: The global challenges of energy access and climate change cannot be solved by any one country or company. Achieving widespread adoption of AI-driven electrolysis and hydrogen energy solutions requires international collaboration. Global agreements, such as the Paris Climate Agreement, must be expanded to include cooperation on the development and distribution of hydrogen-based energy technologies.

2. **Addressing Economic Disparities**: Many developing countries face significant economic barriers to adopting new technologies, especially those requiring substantial infrastructure investments. Ensuring that AI-powered hydrogen systems can be deployed in these regions requires innovative funding mechanisms, international investment, and low-cost, scalable solutions that meet the needs of underserved populations.

3. **Education and Capacity Building**: A sustainable future for AI-driven electrolysis also depends on educating the next generation of leaders, researchers, and workers. By prioritizing education and capacity-building initiatives in the fields of AI, renewable energy, and bioengineering, we can ensure that people around the world are prepared to take advantage of these emerging technologies. Training programs should be designed to provide skills in both the technological and ethical aspects of AI and energy systems, preparing a diverse workforce capable of navigating the complexities of these industries.

Conclusion: Shaping the Future Responsibly

As AI-driven electrolysis continues to develop, the ethical and social responsibilities that accompany these technologies must remain a primary focus. Innovators, policymakers, and stakeholders across the globe must work together to ensure that the benefits of AI-powered hydrogen energy are distributed equitably, that individual rights are protected, and that the technology is deployed with the utmost respect for humanity and the environment.

By embracing a commitment to transparency, accountability, and fairness, we can ensure that AI-driven electrolysis systems serve as a force for positive change, not only revolutionizing energy systems but also improving lives and advancing sustainability for generations to come. The future of energy and human potential is within our grasp—guided by ethical principles and a collective vision for a brighter, more sustainable tomorrow.

Chapter 20: Conclusion: A New Age of Human Potential

As we stand on the precipice of a new era in energy and technology, AI-driven electrolysis represents not only a technical breakthrough but a paradigm shift that can reshape the world. With the potential to harness clean, renewable energy from water and integrate this energy directly into human and societal systems, this technology offers unprecedented opportunities for sustainability, health, and the environment. However, as with any transformative technology, the journey is as much about the ethical considerations and societal impacts as it is about scientific and engineering advances.

This final chapter brings together the themes discussed throughout this book, summarizing the possibilities, challenges, and responsibilities associated with AI-driven electrolysis. It highlights the critical role that innovation, collaboration, and social responsibility will play in unlocking the full potential of this revolutionary technology.

The Transformative Power of AI-Driven Electrolysis

At its core, AI-driven electrolysis is a tool to address two of the most pressing challenges facing humanity: the global energy crisis and the need for sustainable, clean energy solutions. Hydrogen, as a clean fuel, is a key player in decarbonizing industries, transportation, and power generation. By utilizing electrolysis to generate hydrogen from water—using renewable energy sources such as solar and wind—and optimizing this process through artificial intelligence, we are poised to create an energy system that is both sustainable and resilient.

1. Revolutionizing Energy Production: AI's role in optimizing electrolysis systems ensures that hydrogen can be produced efficiently, stored, and utilized. With the ability to predict energy demand and adjust the electrolysis process accordingly, AI can help minimize energy waste, reduce costs, and make hydrogen production a viable alternative to fossil fuels on a large scale. The integration of AI with renewable energy sources also promises to improve grid stability and reliability, enabling decentralized, clean energy production that is adaptable to regional needs.

2. Human-Centered Energy Systems: Beyond large-scale applications, AI-driven electrolysis has the potential to revolutionize healthcare through bio-energy systems that can augment human capabilities. Implantable devices powered by hydrogen fuel cells, for example, could provide self-sustaining, long-term power sources for medical devices, eliminating the need for traditional battery replacements. The application of AI to bioengineering systems allows us to envision a future where the human body and technology work together seamlessly, opening new possibilities for medical advancements and human optimization.

3. Addressing Global Sustainability: The convergence of AI, electrolysis, and hydrogen technologies paves the way for the circular economy, where waste is minimized, resources are efficiently used, and carbon emissions are significantly reduced. By integrating AI-enhanced electrolysis with waste-to-energy technologies and sustainable practices, we can develop systems that are not only energy-efficient but also help restore ecosystems and improve resource management. As hydrogen becomes a key component of the global energy mix, it could lead to a zero-carbon economy where industries operate cleanly, and society functions sustainably.

Challenges Ahead: Technological, Economic, and Ethical Barriers

While the potential for AI-driven electrolysis is vast, significant challenges remain before these technologies can be scaled and widely adopted. The most pressing challenges include:

1. Technological Limitations: Improving the efficiency of electrolysis remains a primary focus for researchers. AI can optimize the electrolysis process, but further advances in materials science, energy storage, and infrastructure development are essential to make hydrogen production truly cost-competitive with fossil fuels. Additionally, breakthroughs in storage and transportation systems are necessary to ensure that hydrogen can be produced and distributed globally without incurring prohibitive costs.

2. Economic Barriers: Despite the long-term potential of AI-driven electrolysis, the initial costs of developing and scaling this technology remain high. Investments in research, infrastructure, and manufacturing are needed to bring down costs and make hydrogen a viable option for all economies, particularly in developing countries where access to clean energy is limited. Economic models and financing solutions that encourage private and public sector investments will be crucial in overcoming these barriers.

3. Ethical and Social Responsibility: As AI-driven electrolysis becomes more integrated into human and industrial systems, it will be vital to ensure that the technology is developed with ethical considerations in mind. Issues of privacy, data security, human rights, and the potential for misuse must be addressed. Additionally, the deployment of this technology should aim to reduce global inequalities by ensuring that the benefits of hydrogen energy are shared equitably across all populations. Access to clean energy must not be limited to affluent countries but extended to those who need it most.

The Role of Collaboration in Shaping the Future

The success of AI-driven electrolysis systems depends on collaboration—across industries, disciplines, and borders. This technology cannot be developed in isolation; it requires interdisciplinary partnerships and a shared vision for a sustainable future. Researchers, engineers, policymakers, and business leaders must work together to overcome technical challenges, drive innovation, and ensure that AI-driven hydrogen technologies are deployed in a socially responsible and environmentally beneficial way.

1. Interdisciplinary Partnerships: The integration of AI, bioengineering, energy systems, and environmental science requires collaboration between experts from diverse fields. Scientific breakthroughs in electrolysis and AI depend on engineers, biologists, chemists, and data scientists working together. Moreover, healthcare professionals, ethicists, and policy experts must collaborate to ensure that AI applications in human biology are both safe and equitable.

2. Global Cooperation: The transition to a global hydrogen economy will require international cooperation. Countries, particularly those with significant renewable energy resources, must collaborate to establish global standards, share knowledge, and invest in infrastructure. Hydroelectric, wind, and solar power from one region can be integrated into the hydrogen production systems of another, enabling the worldwide distribution of clean hydrogen fuel.

3. Policy, Regulation, and Standards: Clear regulations and policies are needed to guide the development and deployment of AI-driven electrolysis systems. These policies must promote innovation while ensuring safety, equity, and sustainability. International agreements should address not only the technical aspects of hydrogen production and storage but also the social and environmental impacts.

A Sustainable Future Powered by AI-Driven Electrolysis

The convergence of AI, electrolysis, and hydrogen energy has the potential to usher in a new era of human potential. As AI optimizes hydrogen production, human biology is enhanced by bio-energy systems, and industries transition to sustainable practices, we will create a cleaner, more equitable world.

However, this transformation will not occur overnight. It requires continued research, interdisciplinary collaboration, and a commitment to ethical principles. Leaders and innovators must stay focused on the long-term goals of sustainability, equity, and global cooperation while addressing the challenges that lie ahead.

As we look to the future, AI-driven electrolysis holds the key to solving some of humanity's most pressing challenges—energy, climate change, and global inequality. By embracing this technology with a clear sense of purpose and responsibility, we can shape a world where energy is abundant, clean, and accessible to all, creating a new age of human potential and a brighter, more sustainable future for generations to come.

Chapter 21: Looking to the Future: Envisioning the Next Steps for AI-Driven Electrolysis

As we stand at the dawn of the next chapter in energy systems, AI-driven electrolysis holds tremendous potential to change not only how we produce energy, but how we live our daily lives. From human-centered energy systems to global sustainability models, the future of AI-powered electrolysis promises profound transformation in multiple sectors, including healthcare, industry, transportation, and beyond. In this final chapter, we will explore the next steps for this groundbreaking technology, addressing the milestones that must be reached and the evolving role of AI in our energy future.

Projections for the Future of AI, Electrolysis, and Hydrogen Energy

The future of hydrogen as a clean energy source, coupled with the power of AI to optimize production processes, is on the cusp of realization. As renewable energy sources like solar and wind continue to grow, AI-driven electrolysis will play an increasingly important role in balancing energy production and consumption. Hydrogen will serve as both a storage medium and an energy carrier, bridging the gap between intermittent renewable energy generation and demand. Here's what the future holds:

1. **Advanced AI and Electrolysis Systems**: As AI algorithms become more advanced, they will significantly improve the efficiency of electrolysis systems. Expect to see electrolysis cells optimized for lower energy consumption, faster production rates, and longer lifespans, making hydrogen a more competitive energy option compared to fossil fuels.

2. **Hydrogen Economy Growth**: The global shift toward hydrogen will accelerate as countries and industries adopt AI-enhanced electrolysis systems. Hydrogen will become a key part of the energy mix, especially in sectors such as transportation, heavy industry, and power generation. By 2050, hydrogen could make up a significant portion of the world's energy, enabling a cleaner, more sustainable energy economy.

3. **Cost Reduction and Scalability**: As research and development continue, AI-driven electrolysis systems will see significant cost reductions. This will allow hydrogen production to scale rapidly, addressing the growing global demand for clean, affordable energy. The transition to hydrogen will not only help meet climate targets but also create new job opportunities and stimulate economic growth in energy sectors worldwide.

Emerging Trends and Technologies that Could Accelerate Progress

The pace at which AI-driven electrolysis can transform energy systems will depend on several emerging trends and technologies:

1. **AI-Optimized Energy Grids**: The integration of AI into energy grid management will allow for the optimization of both hydrogen production and energy storage. Smart grids will use real-time data to adjust power generation, storage, and distribution, ensuring that renewable energy is captured and used efficiently. AI will play a crucial role in making these systems more flexible, cost-effective, and scalable.

2. **Decentralized Hydrogen Production**: The future will see the decentralization of hydrogen production as AI-powered systems allow for smaller, more efficient electrolysis units that can be installed locally, reducing reliance on centralized power plants. This trend will democratize energy production, enabling communities to generate their own hydrogen fuel from local renewable resources, enhancing energy security and resilience.

3. **AI-Driven Bioengineering**: The intersection of AI, biotechnology, and hydrogen energy could give rise to bioengineering innovations that enhance electrolysis systems. Bio-inspired solutions, such as genetically engineered enzymes or bacteria capable of producing hydrogen more efficiently, could drastically improve the sustainability of electrolysis. The fusion of AI and bioengineering will be a critical factor in advancing these technologies.

4. **Carbon Capture and Utilization (CCU)**: The combination of hydrogen production with carbon capture technologies offers a pathway for reducing greenhouse gas emissions. AI-driven electrolysis can be paired with CCU systems to create a closed-loop energy cycle, where carbon emissions are captured, stored, and repurposed, driving the transition to a carbon-neutral economy.

The Convergence of AI, Bioengineering, and Energy: New Opportunities

The convergence of AI, bioengineering, and hydrogen technologies represents an exciting frontier for future innovation. Here are some of the ways these technologies will converge to create new opportunities:

1. **Human-Integrated Bio-Energy Systems**: The integration of AI and hydrogen energy into human-centered systems, particularly in medical technologies, is a promising avenue. Implantable hydrogen-powered devices, AI-optimized bioengineered systems, and enhanced human energy systems could lead to breakthroughs in personalized healthcare. In this future, AI not only manages energy production but actively augments human capabilities, improving health outcomes and enhancing overall quality of life.

2. **AI-Powered Sustainability Innovations**: As AI enhances the capabilities of hydrogen-based systems, we will see the development of sustainable products and services. AI will optimize materials and manufacturing processes to reduce waste, energy consumption, and emissions. The circular economy will thrive, with AI-driven electrolysis playing a central role in creating a sustainable ecosystem where energy, materials, and resources are reused efficiently.

3. **Smart Cities Powered by AI and Hydrogen**: AI and hydrogen have the potential to revolutionize urban infrastructure. Smart cities will integrate AI-driven hydrogen systems into their energy, transportation, and waste management frameworks, creating highly efficient, low-carbon urban environments. AI will manage energy use across buildings, transportation networks, and industries, ensuring that hydrogen is produced and consumed in the most sustainable manner possible.

4. **Artificial Photosynthesis and AI**: Artificial photosynthesis, inspired by the natural process of photosynthesis in plants, holds the potential to provide another source of sustainable hydrogen production. AI could optimize artificial photosynthesis systems to improve efficiency and make them commercially viable. This technology could provide an alternative to conventional electrolysis, allowing for hydrogen production that directly captures solar energy.

Scaling Human-Centric Systems for Widespread Use

For AI-driven electrolysis systems to make a global impact, we must overcome several challenges related to scaling. These challenges are not merely technical—they involve socio-economic considerations, global policy coordination, and infrastructure development:

1. **International Collaboration**: Governments, private companies, and research institutions must work together to accelerate the development and deployment of AI-driven electrolysis systems. This collaboration must be underpinned by a global framework for regulating hydrogen energy, ensuring that systems are deployed responsibly and equitably.

2. **Investment in Infrastructure**: The widespread adoption of hydrogen as a clean energy source requires significant investment in infrastructure. This includes building hydrogen production facilities, storage units, and distribution networks. AI will play a crucial role in optimizing these infrastructure systems, but initial investments are essential to laying the groundwork for the hydrogen economy.

3. **Regulation and Policy Development**: Strong regulatory frameworks will be required to govern the use of AI in energy production, storage, and distribution. Governments must prioritize policies that encourage innovation, ensure safety, and protect the environment. Standards for hydrogen production, AI-powered systems, and bioengineering must be established to guarantee that these technologies are deployed safely and effectively.

Shaping a Sustainable Future Through AI-Driven Electrolysis

As we look ahead, AI-driven electrolysis represents a transformative force that could revolutionize how we think about energy, sustainability, and human potential. By integrating AI with electrolysis to produce clean hydrogen from water, we can address some of the world's most pressing challenges—energy security, climate change, and the need for equitable access to energy resources.

The convergence of AI, renewable energy, bioengineering, and hydrogen technology promises not only a cleaner energy future but a more equitable and sustainable world. By embracing the full potential of AI-driven electrolysis, we can create a new energy paradigm—one that empowers individuals, communities, industries, and nations to thrive in harmony with the planet.

As we move forward, it will be critical to ensure that we harness the power of AI responsibly and ethically, with the goal of improving human lives, restoring ecosystems, and achieving a truly sustainable, low-carbon future. The opportunities are immense, and with thoughtful innovation, collaboration, and dedication, we can make this vision a reality for generations to come.

Chapter 22: Looking Beyond AI-Driven Electrolysis: The Path to the Future

As we explore the future of AI-driven electrolysis, it's important to recognize that this technology, while revolutionary, is not a standalone solution. Rather, it is part of a broader transformation in how we produce, store, and utilize energy on a global scale. The combination of artificial intelligence, hydrogen energy, and electrolysis will play a key role in shaping the future of sustainable energy systems, but the integration of this technology into our global infrastructure will require innovation, collaboration, and new thinking across all sectors.

In this chapter, we look beyond the immediate applications of AI-driven electrolysis, examining the broader future prospects for this technology and the key next steps required to ensure its success.

Pushing the Boundaries of Clean Energy

The global demand for clean energy continues to rise as nations and industries recognize the urgency of addressing climate change. AI-driven electrolysis presents a promising solution to this demand by enabling the production of hydrogen as a clean, scalable energy source. However, we must push the boundaries of what is currently possible with electrolysis and AI to achieve the scale needed to impact the world on a global level.

1. **Scaling Up Electrolysis Technology**: One of the key challenges in harnessing the full potential of AI-driven electrolysis is scaling it to meet the demands of global energy production. Currently, electrolysis systems, while effective, are not yet sufficiently efficient or affordable for widespread deployment. To truly transform the global energy landscape, we need to make further technological advancements to enhance efficiency and lower the cost of electrolysis systems. AI can play a central role in this process by optimizing both the electrolysis process and the systems that support it, from energy generation to distribution.

2. **Integrating Hydrogen into Global Energy Systems**: Hydrogen must become a central player in the global energy grid, and for this to happen, AI-driven electrolysis must be integrated into the broader energy infrastructure. Smart grids powered by AI will be essential in managing the decentralized production and storage of hydrogen energy. By seamlessly integrating renewable sources like wind and solar with hydrogen production, energy grids will be more adaptable to fluctuations in supply and demand, creating a more resilient energy system.

3. **Energy Storage and Distribution Systems**: Beyond the production of hydrogen, efficient storage and distribution systems are critical to ensuring that hydrogen can be reliably used as an energy source. AI-driven electrolysis systems will need to be paired with advanced hydrogen storage technologies that allow for safe, long-term storage of energy. AI can also optimize the logistics of hydrogen distribution, ensuring that it reaches the areas of greatest need while minimizing waste and inefficiency.

The Role of AI in Shaping Future Innovations

AI will not only optimize electrolysis processes but also drive the development of new energy technologies that could complement or enhance hydrogen production. From AI-assisted bioengineering to the optimization of alternative energy sources, the future of clean energy will be shaped by AI innovations.

1. **AI and Bioengineering**: The convergence of AI with bioengineering is one of the most exciting prospects in the future of energy. Advances in artificial photosynthesis and bio-inspired hydrogen production could radically improve the efficiency of electrolysis. AI models will be crucial in designing these bio-based systems, improving their performance, and enabling their integration into existing energy systems. Researchers may also explore genetic engineering to enhance microorganisms that can produce hydrogen directly from organic matter, further revolutionizing clean energy production.

2. **AI in Smart Energy Networks**: The future of energy lies in decentralized, highly flexible networks that can adapt to real-time energy demand. AI will play an essential role in managing these networks, optimizing the flow of energy from production to consumption. Through machine learning algorithms, AI can predict energy demand, anticipate potential issues in energy supply, and automatically adjust the systems to ensure reliability and efficiency. By utilizing AI-driven predictive models, energy grids will be able to handle variable renewable energy sources more effectively, making clean energy more reliable than ever before.

3. **Advanced AI Algorithms for Energy Management**: Machine learning and advanced algorithms will allow for continuous improvement in energy management systems. AI can be used to analyze vast amounts of data to identify patterns and inefficiencies in energy use, helping industries, homes, and governments optimize energy consumption. From automating building energy systems to adjusting power grids for peak demand, AI algorithms will be integral in making our energy use smarter and more sustainable.

Expanding the Impact of AI-Driven Electrolysis on Society

AI-driven electrolysis systems hold the promise of not only addressing the energy crisis but also reshaping industries, economies, and society at large. The impact of these technologies will stretch far beyond the realm of energy production, with the potential to create new jobs, stimulate economic growth, and foster environmental sustainability.

1. **Job Creation in New Energy Sectors**: The growth of the hydrogen economy and the widespread deployment of AI-driven electrolysis will create new job opportunities across multiple sectors. From research and development in AI and electrolysis technologies to jobs in the installation and maintenance of hydrogen production systems, there will be a growing demand for skilled workers in these fields. The transition to a clean energy economy will also create opportunities for workers to retrain in renewable energy technologies, fostering economic growth and technological innovation in both developed and developing countries.

2. **Decentralized Energy Production**: AI-driven electrolysis allows for decentralized energy production, which could empower local communities to produce their own clean energy. This decentralized approach will reduce the reliance on large-scale power plants and give communities more control over their energy sources. By enabling individuals and communities to generate and use hydrogen locally, we can reduce energy poverty and provide more equitable access to clean energy around the world.

3. **Environmental and Societal Impacts**: Beyond energy production, AI-driven electrolysis has the potential to significantly reduce global carbon emissions and help mitigate climate change. By replacing fossil fuels with clean, hydrogen-based energy, we can decrease air pollution, reduce greenhouse gas emissions, and slow the effects of global warming. Furthermore, the circular economy that hydrogen and AI can support will ensure that resources are used more efficiently, reducing waste and encouraging sustainable practices.

The Road Ahead: Collaborating for a Sustainable Future

As we look to the future of AI-driven electrolysis, it is clear that collaboration will be key to realizing its full potential. Governments, industries, and academic institutions must work together to address the challenges of scaling these technologies, building infrastructure, and ensuring equitable access to clean energy. Multilateral cooperation will also be crucial to ensure that the benefits of AI-powered hydrogen production are shared globally.

1. **Global Collaboration for Hydrogen Infrastructure**: The widespread adoption of hydrogen as a global energy source will require unprecedented levels of international collaboration. Governments and international bodies must work together to create regulations, share knowledge, and fund large-scale projects that can accelerate the transition to a hydrogen-based economy.

2. **Cross-Sector Partnerships**: Collaboration between energy providers, technology companies, and governments will be essential in driving the next wave of innovation. From AI development to hydrogen infrastructure, companies must work together to create sustainable solutions that benefit everyone. Public-private partnerships can provide the investment and expertise needed to develop and deploy AI-driven electrolysis technologies on a global scale.

Conclusion: A Vision for the Future

The potential of AI-driven electrolysis is vast, and the future of hydrogen as a clean, sustainable energy source is bright. By continuing to innovate, collaborate, and push the boundaries of what is possible, we can build an energy system that supports human potential, environmental sustainability, and economic growth. The key to success lies in harnessing the power of AI, leveraging the strengths of hydrogen, and working together to create a cleaner, more resilient world.

As we move forward into this new era, AI-driven electrolysis stands as a testament to the transformative power of technology when used with vision, responsibility, and a commitment to the greater good.

Chapter 23: The Convergence of AI, Bioengineering, and Hydrogen: Unlocking Human Potential

The future of AI-driven electrolysis is a convergence of technology, science, and human potential. By harnessing the power of artificial intelligence (AI), bioengineering, and hydrogen energy, we are opening up a new frontier in both energy systems and human augmentation. This chapter delves into the symbiotic relationship between these fields, exploring how they can work together to not only solve global energy crises but also unlock new possibilities for human life, health, and performance.

AI and Bioengineering: A Synergy for Human Augmentation

One of the most compelling aspects of AI-driven electrolysis is its potential to integrate into human systems, particularly through the field of bioengineering. Bioengineering seeks to bridge the gap between biological systems and mechanical or technological enhancements. AI has the capability to significantly improve these bioengineered systems, making them more efficient and capable of producing, storing, and utilizing energy within the human body itself. This fusion of AI and bioengineering could lead to significant advancements in both medical technology and human energy systems.

1. **AI-Powered Bioengineering for Energy Generation**: Imagine a future where the human body can generate its own energy from within, utilizing AI-driven electrolysis processes embedded at a cellular level. Bioengineers are exploring the possibility of embedding nanoscale electrolysis cells inside the body that could harness the body's natural processes to generate hydrogen. Through AI optimization, these systems could regulate energy output in real-time, balancing the energy needs of the body while minimizing waste and inefficiency. This system could be used to power bio-compatible devices or augment human physical and cognitive abilities.

2. **Hydrogen-Powered Human Enhancement**: Building on the potential of electrolysis for sustainable hydrogen production, AI could enable the direct integration of hydrogen as a fuel source within the human body. As AI enhances the performance and efficiency of hydrogen-based systems, bioengineered implants or bio-interfaces might one day allow humans to use hydrogen as a renewable energy source to power internal processes—potentially even enhancing muscle strength, stamina, and cognitive function. The future of human energy could see the body's cells equipped with the ability to produce and store hydrogen autonomously, providing a self-sustaining form of bio-energy.

The Role of AI in Enhancing Biological Systems

Artificial intelligence has already demonstrated its transformative impact across many industries, from healthcare to manufacturing. Its integration with biological systems, however, opens up new realms of possibility—especially in creating more efficient, bio-compatible energy systems. As AI continues to develop, it will be crucial in managing and improving the bioengineering processes that allow for human augmentation.

1. **AI for Personalized Energy Systems**: AI's ability to process and analyze vast amounts of data in real-time will allow for personalized energy systems, optimizing individual energy production based on a person's genetic makeup, health status, and activity levels. This personalized approach could maximize the efficiency of energy generation through electrolysis, ensuring that the body's energy needs are met sustainably and naturally.

2. **AI-Assisted Healthcare Devices**: In healthcare, AI-driven electrolysis systems could power medical devices embedded within the human body. From pacemakers to insulin pumps, devices that require energy to operate could be powered by hydrogen generated through electrolysis, eliminating the need for battery replacements and increasing reliability. AI would continuously monitor the health of the system, adjusting energy levels and ensuring that the medical device operates at peak efficiency. This integration could also extend to AI-driven implants or prosthetics that respond to neural signals, offering more seamless and efficient human augmentation.

3. **Regenerative Medicine**: AI's potential in regenerative medicine is also significant, particularly when combined with bioengineering. For example, AI could optimize the growth of tissues or organs that incorporate AI-driven electrolysis processes to provide constant, low-energy operation within the body. With the ability to regenerate tissue or organs powered by hydrogen, we could potentially extend human longevity and quality of life while reducing the reliance on traditional medical interventions.

AI-Driven Electrolysis and Human Sustainability

Hydrogen, when produced through electrolysis, offers an almost unlimited source of clean energy, and its role in the human body could extend beyond merely powering implants and devices. AI-driven electrolysis could also support a new vision of human sustainability, where individuals are less reliant on external energy sources and can, instead, generate their own energy—making the human body part of a broader energy system.

1. **Self-Sustaining Bio-Energy Systems**: By incorporating AI-enhanced electrolysis at the cellular level, it may be possible to create a closed-loop bio-energy system in which the body not only generates energy from food and oxygen but also directly powers itself using hydrogen. The system would be capable of adjusting energy production based on metabolic needs, maintaining homeostasis without external intervention. This bio-energy cycle could fundamentally transform how we think about human energy and sustainability, creating a model for future healthcare and performance enhancement.

2. **Reducing Environmental Footprint**: As AI-driven electrolysis makes its way into human bio-energy systems, it also opens the door to broader societal impacts. By reducing reliance on external energy systems and creating self-sustaining bio-energy technologies, individuals could drastically lower their environmental footprint. These technologies would also serve as a model for other sectors, with the ability to produce clean hydrogen locally and with minimal environmental impact. This could lead to greater sustainability across industries, including transportation, manufacturing, and agriculture.

Ethical and Social Considerations of Human Augmentation

While the integration of AI-driven electrolysis into human biology offers vast potential for improvement in health, energy systems, and sustainability, it also brings with it significant ethical and social challenges.

1. **The Ethics of Human Enhancement**: Human augmentation, particularly the integration of AI-driven systems within the body, raises important questions about the nature of human autonomy, consent, and equality. Who controls the technology? What standards of care will be applied? How will these technologies be distributed equitably to ensure they do not become a privilege for the wealthy? Ensuring that these advancements benefit all of humanity will require careful ethical consideration, transparent regulations, and inclusivity in design and deployment.

2. **Privacy and Data Security**: AI-powered bio-energy systems will rely on vast amounts of personal data to optimize energy generation, including genetic and health data. This raises concerns about privacy and data security. How can individuals' data be protected while ensuring that AI systems can perform optimally? Establishing clear standards for data management and transparency will be essential to address these concerns and maintain trust in these technologies.

3. **Social Implications**: The integration of AI into human systems will also have far-reaching social implications. Will we see a divide between those who have access to such technologies and those who do not? What new inequalities might emerge as a result of augmented human capabilities? The future of AI-driven electrolysis in humans must be approached with a focus on fairness, equality, and ensuring that these innovations benefit society as a whole.

Conclusion: The Future of AI-Driven Electrolysis and Human Energy

AI-driven electrolysis holds the potential to reshape the future of energy production, sustainability, and human capabilities. As we explore the intersection of AI, bioengineering, and hydrogen energy, we are not only building a new energy system for the world but creating new ways for individuals to sustain and augment themselves. With the continued development of these technologies, the boundaries of what is possible in energy production, medical devices, and human enhancement are expanding.

Yet, with these advancements come important responsibilities. The promise of AI-powered human augmentation must be matched by rigorous ethical frameworks, careful consideration of societal impacts, and global collaboration to ensure that the future of human energy benefits all of humanity.

In this new age of technological possibility, AI-driven electrolysis stands at the frontier of a sustainable, energy-efficient, and augmented future—one where human potential is truly unlocked.

Chapter 24: Conclusion: A New Age of Human Potential

The convergence of artificial intelligence (AI), electrolysis, and bioengineering offers us an unparalleled opportunity to redefine what is possible in human energy systems. As we stand at the crossroads of an energy revolution, AI-driven electrolysis emerges as a transformative technology that not only promises to address global energy challenges but also to unlock new potentials within the human body itself. This chapter serves as a reflection on the journey thus far, the promises of this technology, and the future challenges that lie ahead.

The Evolution of Energy Systems: A Technological Renaissance

Over the course of history, humanity has relied on a range of energy systems to power industries, economies, and everyday life. From the industrial revolution's reliance on coal to the modern transition towards renewable energy sources like solar and wind, energy production has evolved significantly. Yet, despite these advancements, the world continues to face pressing challenges related to energy sustainability, efficiency, and accessibility. AI-driven electrolysis offers a new paradigm—a system where hydrogen, the most abundant element in the universe, can be utilized as a clean, sustainable, and virtually limitless energy source.

The scientific principles behind electrolysis and hydrogen energy have existed for centuries, but the true potential of these technologies has only begun to be realized with the advent of AI. AI has unlocked unprecedented possibilities, enabling us to optimize electrolysis processes, predict energy needs, and even integrate these energy systems directly into human biology. In this technological renaissance, AI-driven electrolysis has the potential to reshape how we generate, store, and use energy, providing a path toward more sustainable and efficient energy systems across the globe.

AI and Human Augmentation: The Next Frontier

One of the most fascinating aspects of AI-driven electrolysis is its potential to augment human capabilities. The concept of the human body generating and utilizing hydrogen-based energy systems represents a monumental leap in bioengineering. From powering medical devices to potentially enhancing physical and cognitive performance, AI-driven electrolysis systems could create a new class of bio-energy systems. This marks the beginning of human augmentation not just as a speculative notion, but as a practical reality where biology and technology merge seamlessly.

However, with such advancements comes the need to address fundamental ethical, societal, and privacy considerations. The integration of AI into human systems will require a global dialogue on the responsible use of these technologies. How do we ensure that these innovations are accessible to all, rather than becoming tools of inequality? How can we safeguard privacy and consent while collecting sensitive biological data to power AI-driven systems? These are questions that will need to be addressed to ensure that the benefits of this technology are shared equitably and responsibly.

Global Sustainability: From Human Energy to Planetary Impact

AI-driven electrolysis holds the promise of not just transforming individual lives but of revolutionizing global energy systems. The use of hydrogen as a clean energy source can significantly reduce carbon emissions, helping to mitigate climate change and support the transition to a zero-carbon economy. With AI's ability to optimize energy production, storage, and distribution, decentralized hydrogen production can be integrated into smart grids, facilitating a sustainable energy ecosystem that is both resilient and efficient.

AI-powered electrolysis also paves the way for a circular economy where energy waste is minimized and resources are used more efficiently. Through innovations in hydrogen storage, AI can help integrate renewable energy sources such as solar and wind, providing continuous, clean power that addresses the intermittency challenges these sources face. As countries around the world continue to grapple with energy insecurity, AI-driven electrolysis offers a vision of a sustainable future where clean energy is accessible and reliable for all.

Challenges Ahead: The Path to Scalable and Inclusive Solutions

While the potential of AI-driven electrolysis is vast, the path to its widespread adoption is fraught with challenges. From technological hurdles to economic and infrastructural barriers, scaling this technology will require substantial investments in research, development, and collaboration across sectors. The integration of AI-driven systems into human biology, for instance, presents complex bioengineering challenges, such as tissue compatibility and long-term sustainability.

Moreover, the societal impacts of these innovations must be carefully managed. As with all transformative technologies, AI-driven electrolysis could exacerbate existing inequalities if not implemented thoughtfully. Ensuring that the benefits of this technology reach underserved populations, both in terms of energy access and human enhancement, will require deliberate policy decisions and international cooperation.

The Ethical Imperative: Ensuring a Fair and Just Future

With great technological power comes great responsibility. As we integrate AI into human systems and harness hydrogen-based energy, we must ensure that these advancements do not come at the expense of individual rights, privacy, or equity. The responsibility of innovators, policymakers, and society at large is to create frameworks that ensure these technologies are used for the greater good, empowering individuals while protecting human dignity and freedom.

AI-driven electrolysis presents a unique opportunity to bridge the gap between sustainable energy production and human empowerment. However, it is vital that we foster a culture of ethical responsibility in the development and deployment of these systems. This means designing AI-powered solutions that are transparent, inclusive, and accountable, and ensuring that their benefits are distributed equitably across society.

Looking to the Future: Embracing a New Age of Human Potential

The integration of AI, electrolysis, and bioengineering holds incredible promise for the future of humanity. As we continue to explore the intersection of technology and human biology, the possibilities are endless. From improving healthcare outcomes to enhancing human performance, AI-driven electrolysis has the potential to transform not only how we generate and use energy but also how we live, work, and thrive as individuals and as a society.

In closing, we are on the cusp of a new era—one where AI, hydrogen, and human potential converge to create a more sustainable, efficient, and equitable future. The technologies discussed in this book are just the beginning of what promises to be a transformative journey toward unlocking the full potential of human energy and power. As we move forward, it will be up to all of us—scientists, engineers, policymakers, and citizens—to ensure that these innovations are harnessed for the benefit of humanity and the planet.

This is the dawn of a new age of human potential—an age defined by limitless possibilities, boundless innovation, and a shared commitment to creating a better world for all.

Chapter 25: A New Age of Human Energy: The Future Awaits

As we stand on the brink of a new era in energy production and human enhancement, the convergence of artificial intelligence (AI) and electrolysis offers immense promise for transforming both our energy systems and our very understanding of human potential. This chapter explores how these advancements can shape our future, offering profound implications not just for energy, but for the way we live, work, and interact with the world around us.

Energy Systems Reimagined: The Role of Hydrogen and AI

For centuries, humanity has relied on fossil fuels, coal, nuclear, and renewable energy sources like solar and wind. Yet, these energy systems still struggle with challenges of sustainability, efficiency, and reliability. Hydrogen energy, unlocked through AI-driven electrolysis, has the potential to change this narrative by providing a clean, abundant, and sustainable alternative. With water as the raw material, hydrogen can be produced virtually anywhere and, when used as a fuel, emits only water vapor as a byproduct.

AI plays a central role in optimizing this process. By enhancing the efficiency of electrolysis systems, predicting energy demand, and automating energy distribution, AI-driven hydrogen production offers a solution that could revolutionize the global energy market. As we incorporate AI into hydrogen energy systems, it is not just the technology itself that transforms but also the way energy is generated, stored, and consumed—ultimately leading to a more decentralized, resilient, and sustainable energy grid.

Human Augmentation and Bio-Energy Systems

Beyond reshaping the global energy landscape, AI-driven electrolysis holds exciting possibilities for human augmentation. Imagine a future where human energy generation is no longer limited to biological processes but enhanced through AI-assisted bioengineering. This could lead to implantable bio-energy systems that provide power for medical devices, prosthetics, or even enhance cognitive and physical abilities. Through bioengineering, electrolysis could enable the human body to harness energy from water, potentially making us self-sustaining at a bio-energy level.

However, this technology raises critical ethical and social concerns. How do we regulate and ensure safety in AI-powered human augmentation? How can we ensure equitable access to these life-enhancing technologies, particularly for marginalized communities? These are questions that will need to be addressed by governments, researchers, and ethicists as AI continues to integrate into human biology.

The Path to Global Sustainability

One of the most promising aspects of AI-driven electrolysis is its potential to address the global sustainability challenge. With renewable energy sources like wind and solar often facing issues of intermittency, hydrogen can serve as a powerful energy storage solution, enabling energy produced during peak hours to be stored and used when needed. By integrating AI to optimize hydrogen production and storage, we can create a circular energy system where clean energy flows seamlessly from production to consumption.

AI-driven electrolysis also paves the way for a truly sustainable global economy. As industries, governments, and businesses work toward achieving carbon neutrality, hydrogen-powered systems could provide the key to reducing emissions across sectors such as transportation, manufacturing, and agriculture. With the proper infrastructure, we could see a future where carbon emissions are dramatically reduced, and the transition to renewable energy becomes the global norm.

Challenges on the Horizon

Despite its incredible potential, the road to large-scale adoption of AI-driven electrolysis and hydrogen energy systems is not without its challenges. Technological limitations, including the need for efficient, low-cost electrolysis systems and storage solutions, must be overcome. Additionally, the economic and logistical hurdles of building a global hydrogen infrastructure are substantial. International collaboration will be essential to address these challenges, and investments in research and development will play a pivotal role in making this future a reality.

Equally important are the societal and ethical considerations. As we integrate AI into human systems and scale hydrogen energy systems, we must be mindful of potential risks—ranging from privacy and safety concerns to unintended consequences on employment and wealth distribution. There is a need for clear, robust governance frameworks to ensure these technologies are used responsibly and equitably.

The Role of Innovation and Education

The future of AI-driven electrolysis and hydrogen energy hinges on continuous innovation and education. To fully realize the potential of these technologies, we must foster a culture of collaboration between industries, academia, and governments. Interdisciplinary research, combining AI, engineering, bioenergy, and environmental sciences, will be key to driving the next wave of breakthroughs.

Education also plays a critical role. As AI and bioengineering continue to evolve, the workforce of tomorrow must be equipped with the skills to work with these advanced technologies. Preparing future generations for this shift will require an emphasis on STEM (science, technology, engineering, and mathematics) education, as well as a broader understanding of sustainability and the ethical implications of emerging technologies.

A Vision for the Future: Limitless Potential

As we close this book, it is clear that AI-driven electrolysis represents more than just a technological advancement. It offers a vision of a future where humanity is no longer bound by the constraints of traditional energy systems. A future where we harness the power of water, AI, and bioengineering to not only create a sustainable energy ecosystem but also augment human potential, paving the way for an era of innovation, empowerment, and global cooperation.

This is just the beginning. The possibilities that lie ahead are boundless, and as we continue to innovate and adapt, we will uncover new solutions to the pressing challenges of energy, sustainability, and human augmentation. AI-driven electrolysis may one day become the cornerstone of a new energy revolution, transforming not just how we live, but how we thrive as individuals and as a global community. The future is now, and it is ours to shape.

This new age of human potential is waiting to be unlocked—through the fusion of AI, hydrogen, and human innovation. Together, we can create a future where energy is not just something we consume, but something we generate, empower, and share across the world. Let us take the first steps toward this sustainable, energetic, and inclusive future.